T0342166

INTEGRATION OF GREEN AND RENEWABLE ENERGY IN ELECTRIC POWER SYSTEMS

INTEGRATION OF GREEN AND RENEWABLE ENERGY IN ELECTRIC POWER SYSTEMS

ALI KEYHANI
MOHAMMAD N. MARWALI
MIN DAI

A John Wiley and Sons, Inc., Publication

Published by John Wiley & Sons, Inc., Hoboken, New Jersey.
Published simultaneously in Canada.

For general information on our other products and services or for technical support, please contact our
Customer Care Department within the United States at (800) 762-2974, outside the United States at (317)
572-3993 or fax (317) 572-4002.

Wiley also publishes its books in a variety of electronic formats. Some content that appears in print may
not be available in electronic format. For more information about Wiley products, visit our web site at
www.wiley.com

Library of Congress Cataloging-in-Publication Data:

Keyhani, Ali, 1942-
  Integration of green and renewable energy in electric power systems / Ali Keyhani,
Mohammad N. Marwali, Min Dai.
    p. cm.
  Includes bibliographical references and index.
  ISBN 978-0-470-18776-0 (cloth)
  1. Distributed generation of electric power.   2. Electric current converters.   3. Renewable energy
sources.   I. Marwali, M.   II. Dai, Min.   III. Title.

TK1006.K44 2010
621.31'21–dc22   2009028767

10 9 8 7 6 5 4 3 2 1

Ali Keyhani dedicates this book to his father:
Dr. Mohammed Hossein Keyhani

CONTENTS

PREFACE

During the past 20 years, Ali Keyhani's research was supported by the National Science Foundation.* This book was conceived based on the work supported by these grants. The authors recognized the need for a book where the three areas of electrical engineering—power system engineering, control systems engineering, and power electronics—must be integrated to address the integration of a green and renewable energy system into electric power systems. The approach to integration of these three areas differs from classical methods. Due to complexity of this task, the authors decided to present the basic concepts and then present a simulation testbed in MATLAB to use these concepts to solve a basic problem in integration of green energy in electric power systems in the form of a project. Therefore, each chapter has three parts: First a problem of integration is stated and its importance is described. Then, the mathematical model of the same problem is formulated. Next, the solution steps are outlined. This step is followed by developing a MATLAB simulation testbed for the same problem. Finally, the results of the project are presented; where applicable, the experimental results are also presented. The book can be used as a textbook for instruction or by researchers. Since every chapter presents a project, an instructor can use these projects with some changes in parameters or control objectives as learning exercises for students. It is suggested that this book be used as an undergraduate and graduate course for students who had some background in power systems, power electronics, and control engineering. However, since the projects of this book are goal-oriented, instructors

*This material is based upon work supported by the first author's National Science Foundation awards ECCS 0501349, ECCS 0105320, and ECCS 0118080. Any opinion, findings, and conclusions or recommendations expressed in this material are those of Ali Keyhani and do not necessarily reflect the views of the National Science Foundation.

can use the book as an interdisciplinary graduate course for electrical and mechanical engineers if the book is supplemented by supporting concepts that students may need.

The book focus is on control of power converters, and we present how the inverters can be controlled to act as steam units and to provide active and reactive power. We will also present the fundamental architectures in design of smart grid distributed generation. In Chapter 1, the two fundamental architectures, namely DC architecture and AC architecture for integration of smart grid distributed generation, are presented. In Chapter 2, we present the inverter control voltage and current in distributed generation systems; in Chapter 3, parallel operation of inverters in distributed generation systems; in Chapter 4, power converter topology for distributed generation systems; in Chapter 5, voltage and current control of a three-phase four-wire distributed generation in island mode; in Chapter 6, power flow control of a single distributed generating unit, in Chapter 7, robust stability analysis of voltage and current control for distributed generation systems; in Chapter 8, PWM rectifier control for three-phase distributed generation system; and in Chapter 9, MATLAB Simulink simulation testbed. In this book, these coordinated control techniques will be presented using a MATLAB Simulink simulation testbed. Throughout the book, we will provide the simulation testbeds used for each chapter in Chapter 9. The instructors can use each chapter to study mathematical modeling with its supporting MATLAB testbed, as well as to provide for students with control projects.

ALI KEYHANI
MOHAMMAD N. MARWALI
MIN DAI

ACKNOWLEDGMENTS

During the past 20 years, Ali Keyhani's research was supported by the National Science Foundation.* This book was conceived based on the work supported by these grants. This book is the application of body of theoretical work of Davison and Ferguson on multivariable robust servomechanism problem (RSP), Francis and Wonham's work on the internal model principle for linear multivariable regulators, and Utkin's work on the sliding mode control technology. Over the years, many graduate and undergraduate students at the Ohio State University have also contributed to the material presented in this book. In particular, Ali Keyhani would like to acknowledge the contribution of the following people: Professor Charles A. Klein, the Associate Chair and Professor H.C. Ko, past Chairman of electrical and computer engineering at the Ohio State University for his support and guidance. The author also wishes to acknowledge the contribution of Mr. Peter Panfil, VP vice president and general manager of Liebert AC Power, Emerson Network Power for many years of collaborative efforts on problem fourmulation and direction of research for control of power converters and his support of the Ohio State University Mechatronics-Green Energy Laboratory and Mr. Jon L. VanDonkelaar, VP New Product Development, Edison Materials Technology Center, for many of collaborative work on many topics for this book. Finally, the authors would like to acknowledge the support of their family.

The individuals at John Wiley & Sons, Inc., involved with book include Paul Petralia, Senior Editor, Michael Christian, Editorial Assistant, and George Telecki, Associate Publisher.

---

*This material is based upon work supported by Ali Keyhani's National Science Foundation awards ECCS 0501349, ECCS 0105320, and ECCS 0118080. Any opinion, findings, and conclusions or recommendations expressed in this material are those of the Ali Keyhani and do not necessarily reflect the views of the National Science Foundation.

CHAPTER 1

SMART GRID DISTRIBUTED GENERATION SYSTEMS

1.1 INTRODUCTION

Energy technologies have a central role in social and economic development at all scales, from household and community to regional, national, and international. Among its welfare effects, energy is closely linked to environmental pollution and degradation, economic development, and quality of living. Today, we are mostly dependent on nonrenewable fossil fuels that have been and will continue to be a major cause of pollution and climate change. Because of these problems and our dwindling supply of petroleum, finding sustainable alternatives is becoming increasingly urgent. Perhaps the greatest challenge in realizing a sustainable future is to develop technology for integration and control of renewable energy sources in smart grid distributed generation.

The smart power grid distributed energy system would provide the platform for the use of renewable sources and adequate emergency power for major metropolitan load centers and would safeguard in preventing the complete blackout of the interconnected power systems due to man-made events and environmental calamity and would provide the ability to break up the interconnected power systems into the cluster smaller regions.

The basic purpose of this book is to introduce the integration and control of renewable energy in electric power systems. Models are important in control of systems because they present the dynamic process of underlying systems. We will present models of green energy systems. These models will be used to develop

Integration of Green and Renewable Energy in Electric Power Systems. By A. Keyhani, M. N. Marwali, and M. Dai
Copyright © 2010 John Wiley & Sons, Inc.

control methods to control the dynamic process of models to accomplish the control objectives.

We present distributed generation (DG) architectures, and then we present the control of converter for utilizing renewable energy sources, such as wind power, solar power, fuel cell (FC) plants, high-speed micro-turbine generator (MTG) plants, and storage devices as local energy sources. This book emphasizes control technology for controlling power converters to supply the loads and to regulate voltage, frequency, and power oscillations. The control technology for the robust global stabilization, tracking, and disturbance attenuation algorithm that are applicable to distributed energy systems will be presented. As part of this objective, we present a MATLAB/ Simulink simulation testbed for presenting the control technology. We will use the time learning approach by introducing the building blocks for analysis and modeling, and then we will present the control technology. We will also present the control methodology to study parallel operation of multiple DG units in low-voltage distribution systems and to mitigate circulating power, the effects of nonlinear loads such as power pre-regulated power-factor-corrected (PFC) loads, voltage and power oscillation due to sudden drop of loads, startup, and loss of local utility. Furthermore, this book will open new vistas for simulation studies and experimental work to address the critical need of industry in expanding the knowledge base in green energy systems, power electronics, and control technology.

Figures 1.1 and 1.2 depict the direct current (DC) architecture and alternating current (AC) architecture of green and renewable power grid DG systems consisting of FC plant, wind turbine, solar arrays, high-speed MTG, and storage systems. The FC and solar power outputs are low-voltage DC that are steps up to a higher-level DC power for processing using DC/DC converters. However, the output power of wind turbines is variable-frequency AC power, and the output power of MTG is high-frequency AC power. For these two sources, the AC/DC or AC/AC converters are used.

In the architecture of Fig. 1.1, the DG sources are connected to a uniform DC bus voltage including the storage system. This will facilitate plug-and-play capability by being able to store the DC power and use DC/AC converters to generate AC power. Today, commercially available storage devices such as flow batteries and battery–flywheel systems can deliver 700 kW for 5 sec to 2 MW for 5 min or 1 MW for up to 30 min, while 28-cell ultra-capacitors can provide up to 12.5 kW for a few seconds. The DG sources of the low-voltage distribution system of Figure 1.3, designated as DGS, is representing a power-generating station that may contain one or all DG sources of Figs. 1.1 and 1.2. These DG units are connected in parallel. The DG system can be operated as an island system or in parallel with the local utility network. In islanding operation the DG system uses the local utility as backup power. First, depending on the availability of the renewable energy sources, the renewable is used to support all or part of the base load, and the remaining DG sources are used to regulate the system voltage and power. However, the island distribution network and its DG sources not only need to be designed to support its own daily load cycle, but also need to be designed with an assumed reliability criterion such as the loss of the largest DG unit. That is, upon occurrence of a large disturbance, the storage devices in conjunction with regulating units are to control

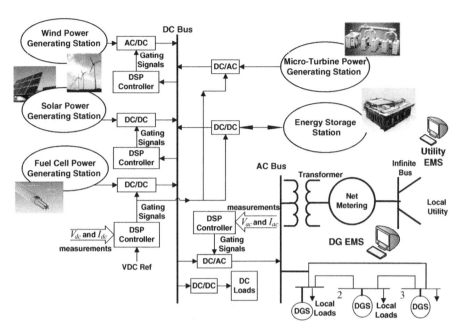

FIGURE 1.1   The DC architecture of green and renewable power grid distributed generation systems.

and to stabilize the low voltage and power oscillations. In the island mode, the stabilization can be achieved using local frequency droop and providing DC power to the DC bus by controlling DC bus voltage and current and charging the storage devices (e.g., battery, flywheel, etc.) as soon as the disturbance is controlled.. To better understand this problem, studying the mix of DG sources with respect to the loss of the largest DG unit in the island network is essential. The proper mix of DG regulating units such as MTG plant (on the order of fraction of seconds), FC plant (on the order of minutes), and storage devices (instantaneously) can be designed to control and to provide their proportional share of power to maintain frequency load regulation and voltage stability. Furthermore, as a last resort, the demand side load management should be used to stabilize the system. However, the system needs to be designed to be sectionalized by switching part of its load to the local utility and continuing to operate as an island with its remaining loads. Another important issue that needs to be investigated is the effects of nonlinear loads that have been increasing their penetration in electric power systems. Today, most loads in hospitals (MRI, CAT scan, etc.) and communication systems (digital signal processing DSP and microcontroller) are pre-regulated PFC. There are many reasons for the PFC technology: (a) The input current waveform is sinusoidal, and hence the injection of current harmonics to the line is very low during the steady state operation. (b) Since the power factor in these types of loads is almost unity, the converters operate at minimum possible operating temperatures. (c) All manufacturers of power converter systems, namely DC power supplies and electric drives, are required to comply with

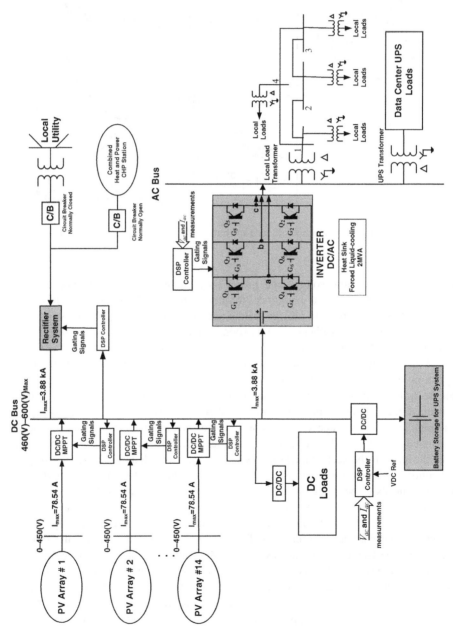

FIGURE 1.2 The DC architecture of a 2-MVA PV station.

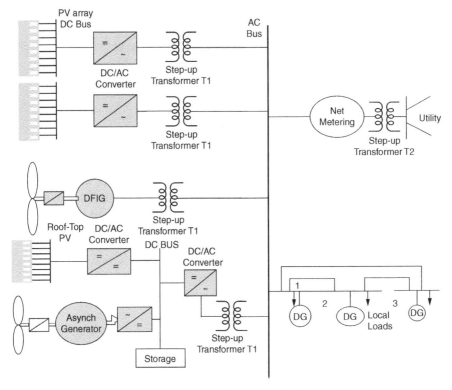

FIGURE 1.3 The Architecture for design of a 2-MVA PV station.

international regulations such as IEC 61000-3-2 and IEEE 519. However, during transient operation or when the power supply system is subjected to a disturbance, such as a drop of loads or the addition of loads or temporary faults, the PFC-type loads would not act as pure resistive loads as they do in their steady-state operations. In fact, these types of loads are highly nonlinear and may act as capacitive and/or inductive loads during disturbances. This type of oscillation has been characterized as bifurcation. The stability study of DGS when they are supplying PFC loads is essential to ensure proper dynamic operation. In-parallel operation with the local utility network creates an important safety issue that needs to be addressed. IEEE standard 1547 spells out DG operation requirements considering the safety issues. However, challenges remain to be addressed in a sudden loss of the utility network.

1.2   DC ARCHITECTURE FOR DESIGN OF A 2-MVA PV STATION

A 2-MVA PV station consists of PV arrays connected in parallel to provide a 2-MVA power output. Each array consists of certain number of PV panels wired in parallel. Therefore the voltage of the PV panel will determine the voltage of each PV array.

TABLE 1.1   Electrical Characteristics of Schott ASE-300-DGF PV Modules

Rated power (P_{max})	300 W
Open circuit voltage (V_{oc})	63.2 V
Maximum power voltage (V_{mp})	50.6 V
Short-circuit current (I_{sc})	6.5 A
Maximum power current (I_{mp})	5.9 A

Source: Affordable Solar website.

A photovoltaic panel is constructed from a number of PV modules wired in series. These modules have certain current–voltage characteristics. Table 1.1 summarizes the electrical characteristics of the discussed PV modules used in this example.

The PV station consists of a DC bus that is connected to the constructed PV arrays. Each array is connected to a DC/DC converter (boost converter) to boost the voltage level to 460 V and a max of 600 V, which is the voltage of the DC bus. The schematic of this PV station is shown in Fig. 1.2.

1.3   PV MODULES

The DC bus voltage is required to be 460 V; however, it can go up to the maximum value of 600 V. According to this requirement, the designed PV array will have a voltage output of 455 V DC at the maximum power rating. Therefore as discussed before, a boost converter will boost the output voltage of the modules to 460–600 V.

The design is based on the ASE ratings and PV array requirements. These ratings are illustrated in Table 1.2. According to the ratings in the table, each PV panel is constructed of 9 "300 DG/50" modules in series. Therefore the output voltage and

TABLE 1.2   ASE Ratings for PV Arrays

Parameter	ASE
Number of arrays	26
Module type	300 DG/50
Modules per array	450
Modules per string	9
Strings per row	2
Power per string STC	2700 W
Design string VOC	595 V
String operating DC	380–430 V
Design array power STC	135 kW
Module failure rate 2004	0.009%

power of each PV panel will be as follows:

$$P_{max} = 300 \times 9 = 2700 \text{ W},$$
$$V_{max} = 50 \times 9 = 450 \text{ V},$$
$$I_{max} = 5.9 \times 9 = 53.1 \text{ A}.$$

Also, each PV array consists of 450 modules; therefore a total of 50 panels should be connected in parallel to construct an array based on ASE ratings. Therefore, the power rating of each PV array will be as follows:

$$P_{max}(\text{Array}) = 135 \text{ (kW)}.$$

The total required power of the PV station is 2 MVA. Based on this requirement, the number of the PV arrays is found.

$$\frac{2000 \text{ kW}}{135 \text{ kW}} \cong 14 \qquad \text{Number of designed PV arrays required for a 2-MVA station.}$$

Each array is connected to a DC/DC converter to boost the voltage level to the maximum.

Therefore, the maximum current under these conditions is

$$I_{max} = 14 \times 5.61 \times 50 = 3.93 \text{ kA}.$$

If we use a boost converter to increase the current, the maximum current out of the converter is found by the energy balance:

$$P_{out} = P_{in} \rightarrow 3.93 \text{ kA} \times 455 = I_{max_{cnv}} \times 460 \rightarrow I_{max_{cnv}} = 3.88 \text{ kA}.$$

According to this calculation, the cable connecting the DC bus to the rest of the system should be rated for a maximum load current of 3.88 kA. Carrying this current at 460 V DC from a PV field to a DC/AC inverter for processing and injection to the utility will result in high power losses. In the process of reducing losses, the DC voltage can be stepped up to higher voltage. This will reduce the power losses but will add to the cost. In addition, the protection of DC system will be a challenge that needs to be resolved.

The results are summarized in Table 1.3.

TABLE 1.3   Number of Modules, Panels, and Arrays in the 2-MVA PV Station Along with ASE

	Number	I_{max} (A)	P_{max} (W)	V_{max} (V)
PV module	6300	5.61	300 W	50
PV panel	700	5.61	2700 W	450
PV array	14	78.54	135 kW	450

1.4 ARCHITECTURE FOR DESIGN OF A 2-MVA PV STATION

Figure 1.3 represented an AC architecture of DG system. For example, a PV system with 2-MW capacity cannot be economical processed at low-voltage DC due to high power losses. The DC system can be used if the DC converters are used to step up the DC voltage of PV system to higher voltages to reduce the power losses. However, today, it is more economical to step up AC voltage to higher voltages for injection to utility system. As shown in Fig. 1.3, the step-up transformer T1 will step up the voltage from the DC/AC converter to a higher voltage. All PV arrays are connected in parallel to the PV system AC bus. In addition, to provide regulating capability for the PV station, a number of PV arrays and wind power energy are processed in DC and the energy is stored in a flow battery or battery–flywheel system. The DC power of storage system is used for regulating the load voltage and load frequency control. The size of the storage system is specified by the regulating requirements of the PV station when it has to operate as an island. The PV station voltage is stepped up with the transformer T2 for parallel operation of the PV station as part of the utility system.

Disturbance due to sudden outage of utility system can cause severe power system stability of the DG system. Furthermore, upon occurrence of faults, the FC and MTG plants fault current could reach a high level, and hence they must be disconnected. Also, after the FC unit is reconnected, the MTG unit may experience instability. Therefore, appropriate control actions need to be taken to stabilize the DG system.

Because of the intermittent nature of renewable energy sources and the slow electrochemical reaction of fuel cells, the need for energy storage devices is inevitable. The energy storage devices will provide operating reserve as fast-acting energy sources when sufficient power can not be provided during load transients and disturbances. Figure 1.4 depicts operation of multiple DG systems connected in parallel in a local network, with the local network connected to the local utility system.

To elaborate on the control methodology, it is essential to recognize that the DG sources, energy storage devices, and the low-voltage network that serve the loads represent multiple entities with conflicting interests. The DG sources are constrained by their dynamic response to disturbances, whether natural or man-made. Upon a major disturbance, the dynamic energy demand may exceed the stored DC bus power reserves, resulting in severe power and voltage oscillation and collapse. Since the loads are mostly power electronics and industrial, their stability margins are relatively close to their stable operating point. Thus, only a finite amount of stored energy in the DC bus (see Fig. 1.1) is available in the DG system. In-depth attention has been given to the needs for storage devices. This problem can be addressed by studying the relationship between energy storage system reserve, demand side management, and the dynamic response of the mix of DG sources in maintaining the DG system stability. MTG plants are fairly fast, and their time responses can change in a fairly wide range. For high-speed MTG the rotor mass is very small; however, since speed is very high, the H constant is in the range of 0.3–1.2; However, there might be other H values out of this range as well. For medium- and low-speed MTG

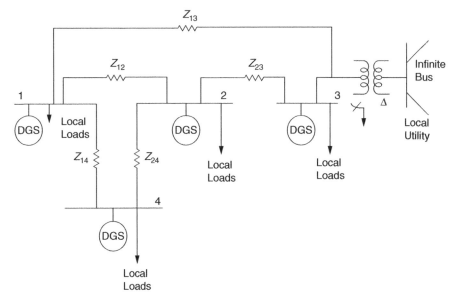

FIGURE 1.4 Low-voltage distributed generation system.

(12,000–50,000 rpm), the *H* constant can be in the range of 0.5–1.5. MTG plants usually use permanent magnet synchronous machines and thus there is no field time constant is associated with them either. FC plants have dynamic response on the order of 5–10 min. Furthermore, the power generated by the renewable generation resources could vary widely, and their operation must be coordinated with amount of DC power storage devices with FC and MTG units that can adjust their generated power in response to disturbances. This coordination is essential for maintaining stable voltage and mitigating the power oscillations. Figures 1.5a and 1.5b depict parallel operation of FC units. Figure 1.6 depicts the control architecture of fuel cells supplied from hydrogen tank. To understand load frequency control and voltage control coordination between various DG sources problems, we need to determine how fast we can control the set points of the MTG and FC plants as soon as the DC bus voltage begins to drop. Furthermore, the use of the advance predictive control can be investigated by using the one-step-ahead load prediction to change the output power of FC and MTG units and to rid through the disturbance and stabilize the DG system. Such a study will also increase understanding the regulating potential of DG system and will minimize the size of DC storage system and demand-side dynamic load-shedding requirements upon disturbances. In fact, we are also trying to determine the mix of DG sources and stored reserve energy. The same problem needs to be investigated when the cluster distributed DG sources lose its connection to the interconnected systems as depicted by Fig. 1.2. Here again, we need to develop control strategy to mitigate the power oscillations and maintain stability. The development of this control system technology is vital to ensure a sustainable

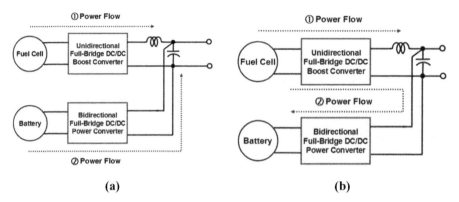

(a) **(b)**

FIGURE 1.5 Block diagram of paralleled-connected FC sources. (**a**) Battery discharge. (**b**) Battery recharge.

network with a significant renewable energy supply contribution. In this book, we use the perfect robust servomechanism problem (RSP). The RSP method had not been applied to practical problems because it requires extensive real-time computations. However, by use of advanced digital signal processors (DSP), this problem has been resolved. We will present this control technology in later chapters. The RSP control guarantees exact asymptotic tracking of the fundamental frequency reference and error regulation of the load disturbance at each of the harmonic frequency included in the servocompensators. The perfect RSP guarantees this property independent of any perturbations in the plant as long as they do not destabilize system. However, it is important to analyze the stability property of the controller under large disturbances and the power-factor-corrected (PFC) loads to ensure proper operation of the converter over its intended operating range. In this book, the stability robustness of the system with the controller will be investigated using structured singular values under nonlinear PFC loads. Specifically, a simulation test bed will be presented for

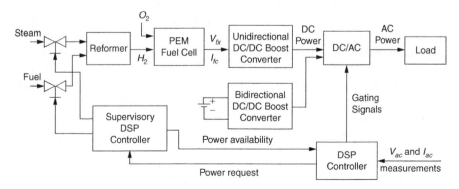

FIGURE 1.6 Fuel cell coordinated control.

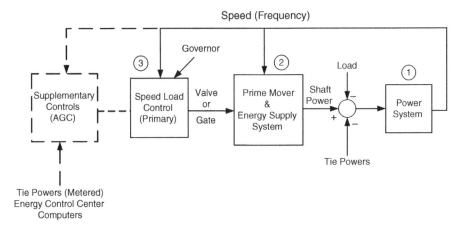

FIGURE 1.7   The automatic generation control of power system (AGC).

the students to study perturbations due to load variations and parameters uncertainties of the system components. A quadratic cost function with separate weighting scalars for plant states and servocompensator states will be used to find solutions to the perfect RSP. Finally, the stability robustness and transient response of the resulting control system will be presented for the systems depicted by Figs. 1.1 and 1.3. Figure 1.7 depicts the architecture for the regulating and control of power disturbances. In this architecture, the primary objective is to control the DC bus voltage as the DC bus voltage drops/increases in response to the fluctuating power demand by the sending control signal to the regulating energy sources to increase/decrease the input power.

The smart grid can be operated in two modes of operations:

1. Synchronized operation with the local utility system.
2. Island mode of operation upon loss of the utility system.

1.5   A DG SYSTEM OPERATING AS PART OF UTILITY POWER SYSTEM

The architecture of Figs. 1.1 and 1.3 satisfy Renewable Portfolio (R/P) laws that have been mandated by many states. This architecture allows selling and buying energy from local utilities. When a DG grid system is connected to a utility, the DG system bus voltage—that is, the infinite bus voltage—is maintained by the local utility. Furthermore, the system frequency is controlled by the power system operator. If the DG inverter has a large capacity in the range of MVA rating, then it will be connected to the subtransmission voltage of the local power system. For example, a system has more than 20,000-MVA capacities feeding its transmission system; if a smart grid DG system is connected to the subtransmission system of this system, the DG system cannot change either the bus voltage or the system frequency. The high-MVA DG grid system will become a part of the utility network, and it will be subjected to power system disturbances. To understand why this is the case, we need to understand the control system that is used by power system operators.

Fig. 1.7 illustrates the energy management system (EMS) of the power system (it is also called the Energy Control Center). The EMS system has two computers that are operating in parallel. Two computers are used to increase the reliability and security of the power system. The function of the energy management system is to control the interconnected system, load and frequency, system voltages, and economic operation of power systems.

The load frequency control system is designed to follow the system load. That is, as load changes—let's say as the load increases (or tie line powers)—the inertia energy stored in the system supplies the deficiency in energy, to achieve a balance between load and generation. This energy is supplied by prime movers (stored energy in rotors). The balance between load and generation must be maintained for a power system to remain stable. When the balance between generation and load is disturbed, the dynamics of generators and loads can cause the system frequency and/or voltages to vary; and if this oscillation persists, it will lead to the system collapse. If the load changes are rapid and the power system frequency drops, then steam units and hydro units control loops will open the hydro gates, and/or turbine steam valves to supply energy to stabilize the system frequency. This action takes place regardless of the cost of energy from generating units. All units that are under load frequency control participate in the regulation of the power system frequency. This is called "governor speed control," as shown in Fig. 1.7. Clearly, the cost of generated energy is not the same for all units. Every 1–2 min, "supplementary control loop," under automatic generation control (AGC), will economically dispatch all units to match load to generation and, at the same time, minimize the total operating cost. Therefore, the AGC will change the set points of generators under its control. This control cycle can be on the order of one minute to several minutes. Figure 1.8 shows the time scale of the power system control.

When a smart grid DG is connected to a utility, then the smart grid DG-generating station operates, using a master-and-slave control technology. The master is the EMS of the utility system. EMS controls the infinite bus voltage and system frequency. It should be observed from Fig. 1.1 that the smart grid DG is connected to the utility system by a transformer. The slave controller controls the AC bus voltage of the inverter (magnitude and phase angle) and inverter current. Therefore, the slave controller of the DG inverter controls active and reactive power. The DG inverter can operate as a unity power factor and leave the voltage control—that is, reactive power (Var) control—to the EMS of the utility system, or it can operate with a leading power factor or a lagging power factor. The digital signal processor (DSP) of the inverter controls the DG inverter. The EMS of the smart grid DG system is interfaced with a DSP controller and a smart net metering system to display the activity of the DG system. If the smart grid is suddenly separated from its local

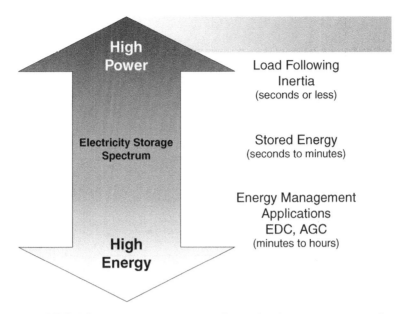

FIGURE 1.8   The energy management time scale of power system control.

utility and the system stability is maintained, then the slave controller takes over load frequency control and voltage control. However, for small rooftop PV sites, the inverter is controlled to produce active power. The VAR support is provided by the local utility. However, for high-MVA generating stations, there will be a purchasing agreement between the utility and DG generating stations on active and reactive power transfer. The word "smart" means that DG generating stations can control their loads and can accept "price signal" and/or "emergency operation signal" from its local utility to adjust their active and reactive power generation. Other designs are also possible. These may include net smart metering communication between the EMS of the local utility and the EMS of DG system. The smart grid DG systems have hardware in place to shed loads, in response to price signal or to rotate nonessential loads and to keep on critical loads. However, since disturbances of a utility system cannot be predicted, it is quite possible that, upon the loss of the utility system, its DG grid systems would be not rapidly disconnected from its utility system, and the stability of DG systems would not be maintained. In this case, DG systems would collapse. However, DG systems can be restored by shedding noncritical loads, and DG systems can return to normal operation in a short time. This problem of stability is a function of how strongly a DG grid system is connected to a utility system. This problem must be studied as system parameters are defined and as voltage level of connecting a DG system to a local utility is defined. Figure 1.9 presents how EMS is controlling load and frequency, system bus voltages, and economic operation of a power system.

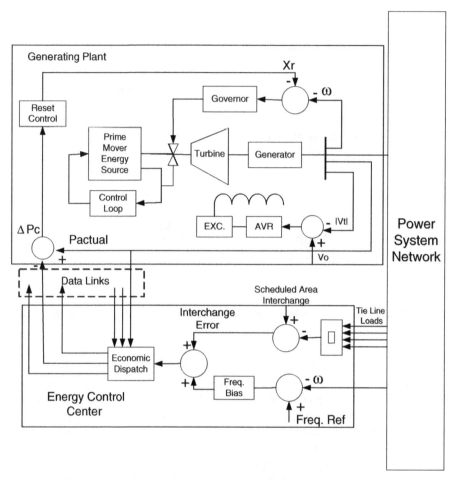

FIGURE 1.9   The energy management system (EMS).

1.6 POWER SYSTEM REACTIVE POWER (VAR) CONTROL

To ensure that power system bus voltages are controlled at their rated values, the power system operators control system Var support. This can be stated as follows: Reactive power generated by generators, plus reactive power generated by capacitors and synchronous condenser, must be equal to reactive power consumed by load, plus system net reactive losses.

To accomplish stable bus voltages, the power system operators control the generator bus voltages by adjusting the field currents of generators. This is done by measuring the terminal voltages of generators and comparing them with set references, as shown in Figs. 1.7 and 1.9. A generator can be operated in three modes by controlling its field current: (1) the overexcited mode, (2) the unity power factor, and

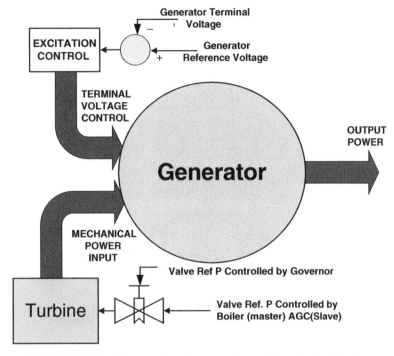

FIGURE 1.10   The operation of a generator as a three-terminal device.

(3) The underexcited mode. The terminal voltages of the generators are set by the system operator to provide active power and reactive power as specified above. By adjusting the generator power factors, the generator's active and the reactive power generation are specified. Then, the turbine valves of the generators are set at a position that will control pressure and steam flow demand, and the positions of the valves correspond to the active power provided by the generators. Therefore, as can be seen in Fig. 1.10, a steam turbine generator is a three-terminal device; that is, the mechanical power is inputted, the generator terminal voltage is controlled, and the generator delivers active power and reactive power.

1.7   AN INVERTER IS ALSO A THREE-TERMINAL DEVICE

Recall that by adjusting the field current, the operator decides on the operating power factor of a generator (that is, active and reactive power production). An inverter can be made to operate in the same three modes and provide active power and reactive power in the same way as a generator.

The operation of the three terminals of an inverter is as follows (Fig. 1.11): (1) input power is supplied from a DC source (battery storage); (2) AC power is supplied to the load; and (3) DSP controller, based on its control technology, computes the switching

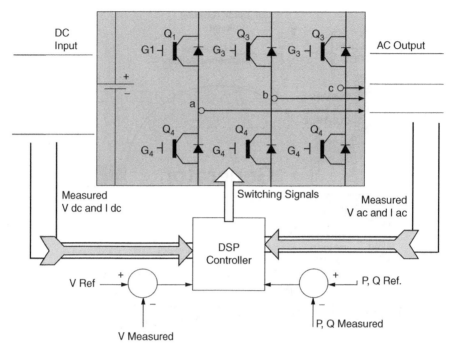

FIGURE 1.11 The operation of an inverter as a three-terminal device.

policy that *controls the phase angle between the terminal voltage and current, that is, the power factor.* Therefore, using the appropriate control technology, we can operate an inverter, such that, it can produce active power and reactive power. The frequency of generated power is set by pulse width modulation (PWM). The details are given in Chapter 2.

In Fig. 1.12, we have multiple control loops. The first control loop controls the DG AC bus voltage, that is, the load bus. The second control loop controls the AC bus current (i.e., the load current). This is accomplished by controlling the phase

Perfect Robust Servo Mechanism Control

FIGURE 1.12 The inverter control system.

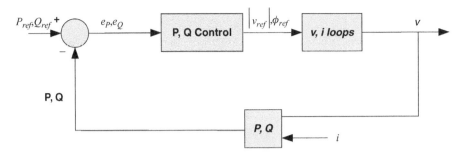

FIGURE 1.13   The grid-connected inverter P and Q control.

angle of the load current with respect to DG AC bus voltage. Finally, the stabilizing compensator eliminate the unwanted harmonics.

In Fig. 1.13, the desired active and reactive power to be injected into the local utility is specified by the P ref and the Q ref. The AC bus voltage and current, the P and Q injected to the local utility, are calculated and compared with the reference active power and reactive power. The P–Q control loop determines the desired control actions for the amount of power to be transferred to the AC bus. Therefore, the inverter is controlled to deliver active power and reactive power per the purchased agreement or based on "price signal" from a power system operator. If a utility system is subjected to a major disturbance such as faults, it can pull down DC bus voltage if the DG system is not rapidly disconnected from the utility. This problem must be investigated by modeling the battery system and developing an equivalent coherent generator model of the utility power network.

1.8   THE SMART GRID PV-UPS DG SYSTEM

When combining the integration of green energy sources with un interruptible power supply (UPS), the architecture presented by Fig. 1.14 will increase the security of the system by attempting to isolate UPS loads from utility disturbances. In this architecture, two power converters are used. First, a bidirectional converter is used to charge the battery system and inject power into the local utility. A second inverter is used to support UPS loads and local smart grid loads.

The architecture depicted by Fig. 1.14 is essentially the same as in the smart grid DG system of Figs. 1.1 and 1.3. However, this architecture operates at three frequencies. The DC bus operates at zero frequency, the portion of the system connected to the local utility operates at synchronized power system frequency, and the AC bus of the UPS system operates at the inverter frequency. The architecture in Fig. 1.14 provides for the isolation of UPS AC bus from utility system as long as the storage system can provide this isolation. However, if the utility system is subjected to large disturbances that result in rapid discharge of the storage system and the drop in DC bus voltage is excessive, this will result in the collapse of AC bus of UPS system. To understand this problem, let us assume that the UPS-PV system is injecting power

18

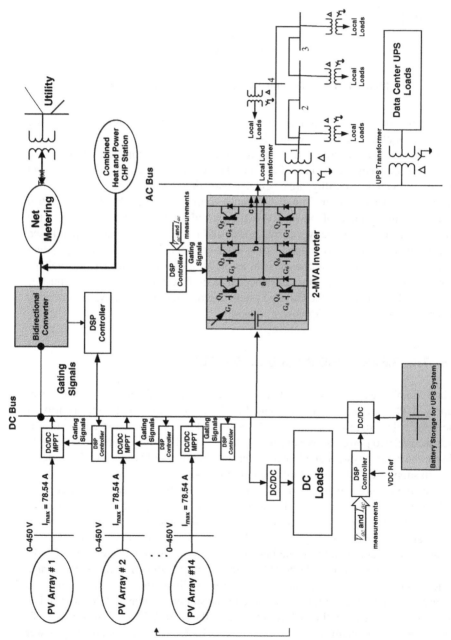

FIGURE 1.14 The smart grid PV-UPS DG system.

into the utility system. Then, unexpectedly, the line that is transferring power into the local utility trips out. Upon this disturbance, the resulting oscillations may result in voltage oscillations of the DC bus and the collapse of the smart grid UPS-PV DG system. The stronger the connection (i.e., the higher the voltage) to the local utility system, the higher the chance that voltage oscillations will penetrate its smart grid UPS-PV. This problem must be studied by modeling the smart grid UPS-PV system and developing a coherent equivalent network for the local utility network in order to determine the protective relay setting for the separation of the smart grid UPS-PV system from the local utility.

1.9 THE SMART GRID SPLIT DC BUS UPS-PV DG SYSTEM

When combining the integration of green energy sources with un interruptible power supply (UPS), the architecture presented in Fig. 1.14 will increase the security of the system by attempting to isolate UPS loads from utility disturbances. However, the stability and the security of UPS loads *cannot* be guaranteed. To isolate the UPS loads, the smart grid DC bus of the UPS-PV DG system is split into two sections as shown in Fig. 1.15 through a normally open circuit breaker. This architecture operates

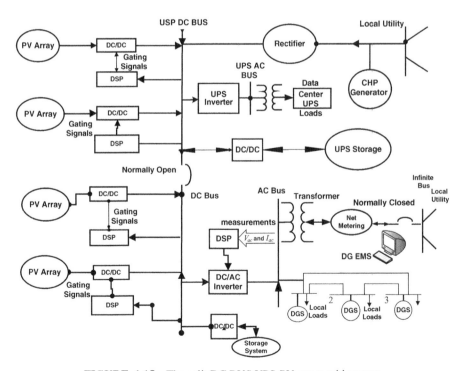

FIGURE 1.15 The split DC BUS UPS-PV smart grid system.

at the following frequencies: (1) The DC bus system operates at zero frequency. (2) The 2-MVA inverter, which connects the DG system to the local utility, operates at the power system frequency (synchronized frequency). (3) Through a separate UPS inverter, the UPS loads are connected to the DG system. The UPS loads operate at the UPS inverter frequency. The split DC bus UPS system would have the same reliability and security as classic UPS technology since it is connected to its utility network, using a rectifier. Therefore, the DC bus voltage of UPS loads is protected from local utility disturbances.

1.10 THE ISLAND MODE OF OPERATION

Upon separation from a utility system, if a smart grid DG stays stable, the inverter of the DG system regulates its frequency. If there is only one inverter in the DG system, load frequency control is accomplished through inverter reference frequency using the Pulse Width Modulation (PWM) technique as part of the DSP controller. The DG system inverter controls its load at the reference AC bus voltage. As load changes, the DSP controller corrects the AC bus voltage. Therefore, as depicted in Figs. 1.1, 1.3, 1.14, and 1.15, the DG inverter follows the load changes and converts DC power to AC, to satisfy active power and reactive power. Again, system load is matched to the generated power from the DG inverter. In short, the DG system is a micro-grid power system that operates like an AC power system.

1.11 THE PARALLEL OPERATION OF INVERTERS

If a smart grid generating station has two and more inverters that are operating in parallel, a new problem would arise. Two inverters in parallel, as shown in Fig. 1.16, would act as two transmission lines with unequal loading, and *circulating zero-sequence current would flow between the parallel inverters*. The circulating current can be excessive when both inverters are sharing nonlinear loads, typically a load consisting entirely or partly of rectifiers—for example, diode or thyristor rectifiers. The sharing of load current harmonics between inverter units is critical. The nonlinear loads cause the distortion of the current supplied by inverters. The load current will have harmonics, in addition to the fundamental frequency. If current harmonics are not distributed evenly between the inverters, there will be a risk of overloading the individual inverter, due to the circulating current that would cause heating. This heating would result in a reduction of lifetime operation of inverters.

This problem must be investigated further as system parameters are defined and the voltage level of the connecting the DG system to its local utility is defined. We will present the operation of parallel inverters in a later chapter.

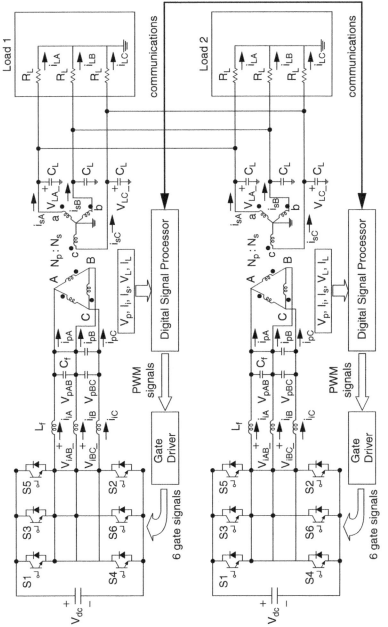

FIGURE 1.16 The parallel operation of two inverters.

1.12 THE INVERTER OPERATING AS STEAM UNIT

If a DG inverter system is designed to operate as a steam unit, then it can participate as part of power system voltage and frequency control. In this case, the DG inverter must have well-defined dynamic response. Such a DG generating station would be very valuable to its local utility, since the excess power can be used as "real-time spinning reserve." If the DG system is designed as a steam power generating system, then it can sell its power to its local utility, based on "price signal," by shedding its own nonessential local loads. If a DG generating system qualifies as provider of real-time spinning reserve, such a DG system would receive payment from its local utility, even if its generated power is not used. But, it must be ready to shed its loads, if needed, and be under the utility EMS to respond to the local utility power system emergency.

The steam power plant has stored energy in its boiler and stored energy in its inertia of rotor. This allows the power system operator to control the frequency of the system. Again, in a power system, the loads are controlled by the end users. The power system operator controls the system frequency to follow the system load demand. This objective is accomplished by comparing the actual system frequency with a reference frequency. Based on the deviation from the set point (i.e., 60 Hz in the United States), EMS opens and closes the turbine valve systems, as shown in Fig. 1.4. However, the opening and closing of turbine valves must follow the boiler's operating condition limits. The drop in steam pressure and steam conditions must not deviate from their set limits, since the drop in steam pressure and temperature will result in water droplets and damage to turbines.

We can make the battery storage system act as a boiler. However, this must be accomplished by following the limits on the discharge rate of battery storage systems, that is, state of charge (SOC). However, as with a boiler, it is necessary to make sure that the battery systems will not be subjected to excessive discharge that damages the battery system or reduces the life of the battery storage system. We can use a battery storage system with a flywheel or a super-charging capacitor to make the combination of a battery–flywheel storage system, with its inverter acting as a steam power plant, as shown in Fig. 1.17. In this architecture, the flywheel system would provide inertia energy rapidly as the DC voltage bus drops; and battery storage system would take over, under appropriate control, to keep the system stable.

1.13 THE PROBLEM OF POWER QUALITY

The total harmonic distortion (THD) is an important consideration in power quality and the operation of DG systems. The total harmonic distortion (THD) of a power signal is a measurement of the power harmonic distortion present and is defined as the ratio of the sum of the powers of all the harmonic components to the power of the fundamental power frequency. The impact of harmonics is their effect on power system components and loads. Transformers are major components in power systems. The increased losses due to harmonic distortion can cause excessive losses and, hence,

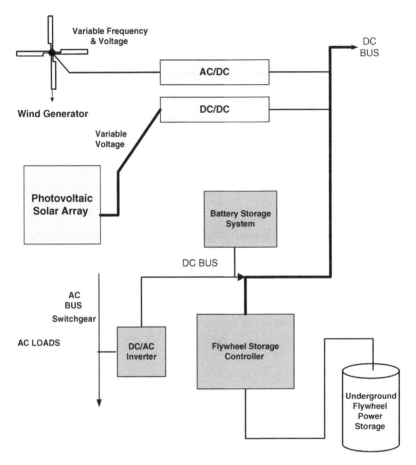

FIGURE 1.17   A DG generating station acting as a steam unit.

abnormal temperature rise. The unwanted harmonics can be eliminated using perfect robust servomechanism control as shown in Fig. 1.12. The control strategy combines the *perfect RSP* controller for low THD output voltages. The discrete sliding mode current controller provides for fast overcurrent protection. In the following chapters, we will present the modeling and control of inverters for integration of renewable and green energy in electric power systems.

The advocated load sharing technique will address the following areas: The technique shall not be sensitive to component mismatches, measurement error, or unbalanced load or wire impedances. We will establish the sharing of harmonic components of the currents without significantly degrading the performance of the output voltages avoiding the existence of a single point of failure in the paralleled units' configuration.

In Fig. 1.18a, the control objective is to control the FC and MTG set points in response to disturbances. In Fig. 1.18b, the control objective is to control DC/DC converters and the DC bus voltage in response to disturbances. Finally, in Fig 1.18c,

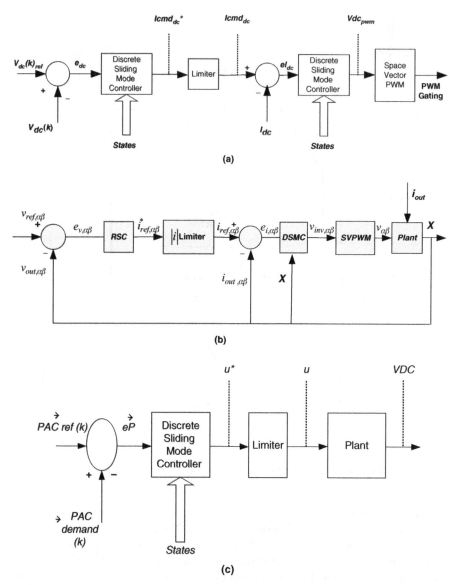

FIGURE 1.18 The block diagram of the proposed coordinated control. (**a**) DC bus voltage control. (**b**) AC bus voltage control. (**c**) Active power control.

the control objectives are to stabilize load voltage and active and reactive power outputs of DC/AC converter in response to disturbances.

The above control technology will be presented in this book. In Chapter 2, we will present the inverter control voltage and current in distributed generation systems; in Chapter 3, parallel operation of inverters in distributed generation systems; in

Chapter 4, power converter topology for distributed generation systems; in Chapter 5, voltage and current control of a three-phase four-wire distributed generation in island mode; in Chapter 6, power flow control of a single distributed generating unit; in Chapter 7, robust stability analysis of voltage and current control for distributed generation systems; in Chapter 8, PWM rectifier control for three-phase distributed generation system; in Chapter 9, modeling of proton exchange membrane fuel cell; and in Chapter 10, MATLAB simulation testbed. In this book, these coordinated control techniques will be presented using a MATLAB simulink simulation test bed. Throughput the book, we will provide the simulation test beds. The instructors can use each chapter mathematical modeling with its supporting MATLAB to provide students with control projects.

CHAPTER 2

INVERTER CONTROL VOLTAGE AND CURRENT IN DISTRIBUTED GENERATION SYSTEMS

2.1 THE POWER CONVERTER SYSTEM

The power converter system used in this chapter consists of a typical three-phase PWM voltage inverter with L–C output filter (*Linv* and *Cinv*) and a delta–wye transformer that act as both a potential transformer and a electrical isolation to the load. Fig. 2.1 shows a circuit diagram of the system. Notice that the delta-wye transformer converts a three-wire (UVW) power system of the inverter to a four-wire (*XYZ-N*) system for the load. Small capacitors (denoted as Cgrass in Fig. 2.1) are added at the load side of the transformer to provide further harmonics filtering and stabilization of the load voltages. A DSP (digital signal processor) system controls the operation of the power converter, providing required PWM gating signals to the power devices. Voltages and currents measured by the DSP system for control purposes are shown labeled in Fig. 2.1. The line-to-neutral load voltages (at points xyz-n in Fig. 2.1) are denoted as $Vload_{an}$, $Vload_{bn}$, and $Vload_{cn}$, the load phase currents are denoted as $Iload_a$, $Iload_b$, and $Iload_c$, the line-to-line inverter filter capacitor voltages (at points uvw in Fig. 2.1) are denoted as $Vinv_{ab}$, $Vinv_{bc}$, and $Vinv_{ca}$, and the inverter phase currents are denoted as $Iinv_a$, $Iinv_b$, and $Iinv_c$.

For development of the control algorithm, a state-space model of the system is needed. Each phase of the delta–wye transformer has been modeled as an ideal transformer with leakage inductance *Ltrans* and series resistance *Rtrans* on the secondary winding as shown in Fig. 2.2. The secondary transformer currents are denoted as $Isnd_a$, $Isnd_b$, and $Isnd_c$.

Integration of Green and Renewable Energy in Electric Power Systems. By A. Keyhani, M. N. Marwali, and M. Dai
Copyright © 2010 John Wiley & Sons, Inc.

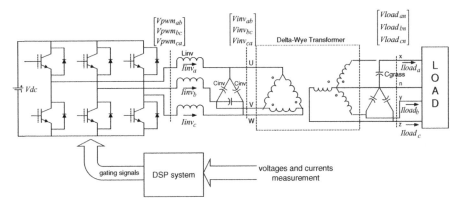

FIGURE 2.1 Power converter system.

Using the transformer model in Fig. 2.2, the dynamic equations of the output filter circuit in Figure 2.1 can be written as in Eqs. (2.1a)–(2.1d):

$$\frac{d\vec{V}inv_{abc}}{dt} = \frac{1}{3 \cdot C_{inv}}\vec{I}inv_{abc} - \frac{1}{3 \cdot C_{inv}}Tr_i \cdot \vec{I}snd_{abc}, \tag{2.1a}$$

$$\frac{d\vec{I}inv_{abc}}{dt} = \frac{1}{L_{inv}}\vec{V}pwm_{abc} - \frac{1}{L_{inv}}\vec{V}inv_{abc}, \tag{2.1b}$$

$$\frac{d\vec{V}load_{abc}}{dt} = \frac{1}{C_{load}}\vec{I}snd_{abc} - \frac{1}{C_{load}}\vec{I}load_{abc}, \tag{2.1c}$$

$$rcl\frac{d\vec{I}snd_{abc}}{dt} = -\frac{R_{trans}}{L_{tran}}\vec{I}_{snd} + \frac{1}{L_{tran}}Tr_v \cdot \vec{V}inv_{abc},$$

$$-\frac{1}{L_{tran}}\vec{V}load_{abc}, \tag{2.1d}$$

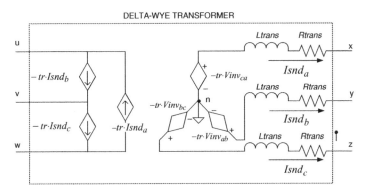

FIGURE 2.2 *Transformer* model.

where the voltages and currents vectors are defined as in Eq. (2.2).

$$\vec{V}inv_{abc} = \begin{bmatrix} Vinv_{ab} & Vinv_{bc} & Vinv_{ca} \end{bmatrix}^T$$

$$\vec{V}load_{abc} = \begin{bmatrix} Vload_a & Vload_b & Vload_c \end{bmatrix}^T,$$

$$\vec{I}load_{abc} = \begin{bmatrix} Iload_a & Iload_b & Iload_c \end{bmatrix}^T \quad \vec{I}snd_{abc} = \begin{bmatrix} Isnd_a & Isnd_b & Isnd_c \end{bmatrix}^T,$$

$$\vec{I}inv_{abc} = \begin{bmatrix} Iinv_{ab} & Iinv_{bc} & Iinv_{ca} \end{bmatrix}^T$$

$$= \begin{bmatrix} Iinv_a - Iinv_b & Iinv_b - Iinv_c & Iinv_c - Iinv_a \end{bmatrix}^T.$$

$$(2.2)$$

Matrices Tr_i and Tr_v in Eqs. (2.1a) and (2.1d) denote the currents and voltages transformations of the delta–wye transformer. Denoting the transformer's turn ratio as tr, these matrices are given by (2.3)

$$Tr_i = tr \cdot \begin{bmatrix} 1 & -2 & 1 \\ 1 & 1 & -2 \\ -2 & 1 & 1 \end{bmatrix}, \quad Tr_v = tr \cdot \begin{bmatrix} 0 & 0 & -1 \\ -1 & 0 & 0 \\ 0 & -1 & 0 \end{bmatrix}. \quad (2.3)$$

To obtain a state-space model of the system, the dynamic equations (2.1) are transformed to the DQ0 stationary reference frame using the transformation

$$\vec{f}_{qd0} = K_S \cdot \vec{f}_{abc}, \quad (2.4)$$

with

$$K_S = \frac{2}{3} \begin{bmatrix} 1 & -0.5 & -0.5 \\ 0 & -\sqrt{3}/2 & \sqrt{3}/2 \\ 0.5 & 0.5 & 0.5 \end{bmatrix},$$

$$f_{qd0} = [f_q, f_d, f_0]^T, \quad f_{abc} = [f_a, f_b, f_c]^T,$$

where $\vec{f}_{abc}$ denotes the abc voltages and currents defined in Eq. (2.2), and $\vec{f}_{qd0}$ denotes the corresponding DQ0 stationary reference frame variables. The circuit dynamics can then be written as in Eqs. (2.5a–2.5d):

$$\frac{d\vec{V}inv_{qd}}{dt} = \frac{1}{3 \cdot C_{inv}} \vec{I}inv_{qd} - \frac{1}{3 \cdot C_{inv}} Tr_{iqd0} \cdot \vec{I}snd_{qd0}, \quad (2.5a)$$

$$\frac{d\vec{I}inv_{qd}}{dt} = \frac{1}{L_{inv}} \vec{V}pwm_{qd} - \frac{1}{L_{inv}} \vec{V}inv_{qd}, \quad (2.5b)$$

$$\frac{d\vec{V}load_{qd0}}{dt} = \frac{1}{C_{load}}\vec{I}snd_{qd0} - \frac{1}{C_{load}}\vec{I}load_{qd0}, \qquad (2.5c)$$

$$\frac{d\vec{I}snd_{qd0}}{dt} = -\frac{R_{tran}}{L_{tran}}\vec{I}snd_{qd0} + \frac{1}{L_{tran}}Trv_{qd} \cdot \vec{V}inv_{qd} - \frac{1}{L_{tran}}\vec{V}load_{qd0}, \qquad (2.5d)$$

where the matrices Tri_{qd0} and Trv_{qd} are defined as

$$Tri_{qd0} = \left[K_s \cdot Tr_i \cdot K_s^{-1}\right]_{row1,2} = tr \cdot \frac{3}{2}\begin{bmatrix} 1 & \sqrt{3} & 0 \\ -\sqrt{3} & 1 & 0 \end{bmatrix}, \qquad (2.6a)$$

$$Trv_{qd} = \left[K_s \cdot Tr_v \cdot K_s^{-1}\right]_{col1,2} = tr \cdot \frac{1}{2}\begin{bmatrix} 1 & -\sqrt{3} \\ \sqrt{3} & 1 \\ 0 & 0 \end{bmatrix}. \qquad (2.6b)$$

Note that, due to the three-wire system of the inverter and filter, the zero components of the inverter voltages ($\vec{V}inv_{qd}$), the inverter currents ($\vec{I}inv_{qd}$), and the input PWM voltages $\vec{V}pwm_{qd}$ are trivial and they do not appear in (2.5a–2.5d).

Using the dynamic equations in (2.5a–2.5d), equivalent circuits can be drawn in the DQ0 stationary reference frame as shown in Fig. 2.3. From Fig. 2.3, it can be seen that the 0-axis equivalent circuit of the load side of the transformer is completely decoupled from the qd-axis equivalent circuits. This shows that the zero components of the load voltages and the secondary transformer currents ($Vload_0$ and $Isnd_0$) are uncontrollable by the input PWM voltages $\vec{V}pwm_{qd}$. A linear controllability analysis performed with Eqs. (2.5a–2.5d) can also be used to show this fact. The 0-component of the load currents will be nonzero under unbalanced load condition or under nonlinear load in the form of triplent harmonics. These 0-components will create 0-component of the load voltages, which is undesirable. Since these 0-components are uncontrollable, no action can be done by the control to compensate them. However, as is apparent from Fig. 2.3, the small capacitor at the output along with the leakage inductance of the transformer act as an L–C filter that can be tuned to provide attenuation to the effect of the 0-component of the load current to the load voltages. From Fig. 2.3, the amount of this attenuation at steady state can be calculated from

$$|Vload_0(j\omega)| = \left|\frac{j\omega \cdot L_{trans} + R_{trans}}{1 - \omega^2 L_{trans}C_{grass} + j\omega \cdot R_{trans}C_{grass}}\right| \cdot |Iload_0(j\omega)|, \qquad (2.7)$$

where $\omega = 2\pi f$ and f is a harmonic frequency of interest.

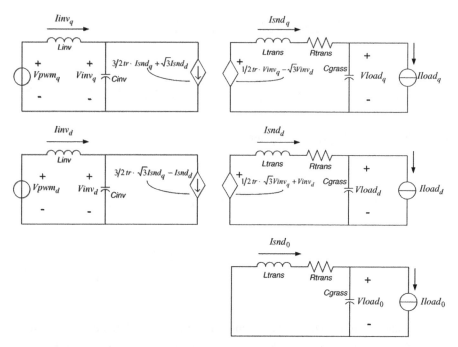

FIGURE 2.3 Equivalent circuits in the DQ0 stationary reference frame.

2.2 CONTROL THEORY

2.2.1 Perfect Control of Robust Servomechanism Problem

Finding a solution to a *robust servomechanism problem (RSP)* for a system involves finding a controller for the system such that: Given a modeled class of unstable tracking/disturbance signals, exact asymptotic tracking/regulation occurs for all plant perturbations which do not produce instability. A *perfect* controller for the *RSP* solves the *RSP* and also provides arbitrarily good transient error, with no unbounded peaking in the error response of the system. Below some of the main results of the *perfect RSP* discussed in literture are summarized.

Consider the plant to be regulated described by the following equations:

$$
\begin{aligned}
\dot{x} &= Ax + Bu + Ed, \\
y &= Cx + Du + Fd, \\
y_m &= C_m x + D_m u + F_m d, \\
e &= y_{ref} - y,
\end{aligned}
\tag{2.8}
$$

where $x \in R^n$, $u \in R^m$ are the inputs, $y \in R^r$ are the outputs to be regulated, $y_m \in R^{rm}$ are the measurable outputs, $d \in R^\delta$ are the disturbances, $y_{ref} \in R^r$ are the reference input signals, and $e \in R^r$ are the error in the system. It is assumed that the disturbance d arises from the following class of systems:

$$\dot{\eta}_1 = \Psi_1 \eta_1, \qquad d = C_1 \eta_1, \qquad \eta_1 \in R^{n1} \tag{2.9}$$

and the reference input signals y_{ref} arise from the following:

$$\dot{\eta}_2 = \Psi_2 \eta_2, \qquad y_{ref} = C_2 \eta_2, \qquad \eta_2 \in R^{n2}, \tag{2.10}$$

with $eig(\Psi_1) \subset C^+$ and $eig(\Psi_2) \subset C^+$, where $eig(\cdot)$ denotes the eigenvalues and C^+ denotes the closed right complex half-plane. This class of signals includes most classes of signals that occur in application problems—for example, constant, polynomial, sinusoidal, polynomial–sinusoidal, and so on. Let $\Lambda := \{\lambda_1, \lambda_2, \ldots, \lambda_p\}$ be the zeros of the least common multiple of minimal polynomial of Ψ_1 and Ψ_2 (multiplicities repeated), then a linear controller that solves the RSP for (2.8) consists of the following structure:

$$u = \xi + K_1 \eta, \tag{2.11}$$

where $\eta \in R^{r \cdot p}$ is the output of a *servocompensator* and $\xi \in R^m$ is the output of a *stabilizing-compensator S*. The servocompensator has the form

$$\dot{\eta} = A_c \eta + B_c e \tag{2.12}$$

with

$$A_c = \text{block diag} \, (\underbrace{\Omega, \ \Omega, \ldots, \Omega}_{r}),$$

$$B_c = \text{block diag} \, (\underbrace{\gamma, \ \gamma, \ldots, \gamma}_{r}),$$

and

$$\Omega = \begin{bmatrix} 0 & 1 & 0 & \cdots & 0 \\ 0 & 0 & 1 & \cdots & 0 \\ \vdots & \vdots & \vdots & \vdots & \vdots \\ -\sigma_1 & -\sigma_2 & -\sigma_3 & \cdots & -\sigma_p \end{bmatrix}, \qquad \gamma = \begin{bmatrix} 0 \\ 0 \\ \vdots \\ 0 \\ 1 \end{bmatrix},$$

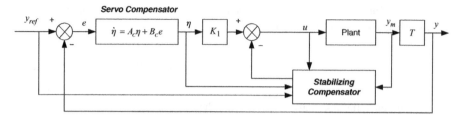

FIGURE 2.4   General robust servomechanism controller.

where the coefficients σ_i, $i = 1, 2, \ldots, p$, are given by the coefficients of the poly-nomial $\prod_{i=1}^{p} (\lambda - \lambda_i)$; that is,

$$\lambda^p + \sigma_p \lambda^{p-1} + \cdots + \sigma_2 \lambda + \sigma_1 := \prod_{i=1}^{p} (\lambda - \lambda_i). \tag{2.13}$$

The servocompensator gain K_1 in Eq. (2.11) and the *stabilizing-compensator S* should be found to stabilize and "give desired behavior" to the following stabilizable and detectable system:

$$\begin{bmatrix} \dot{x} \\ \dot{\eta} \end{bmatrix} = \begin{bmatrix} A & 0 \\ -B_c C & A_c \end{bmatrix} \begin{bmatrix} x \\ \eta \end{bmatrix} + \begin{bmatrix} B \\ -B_c D \end{bmatrix} u,$$

$$\begin{bmatrix} y_m \\ \eta \end{bmatrix} = \begin{bmatrix} C_m & 0 \\ 0 & I \end{bmatrix} \begin{bmatrix} x \\ \eta \end{bmatrix} + \begin{bmatrix} D_m \\ 0 \end{bmatrix} u. \tag{2.14}$$

A general controller that solves the robust servomechanism problem can be dia-grammatically shown as in Fig. 2.4.

Note that, given a modeled class of references/disturbances, the servocompensator is unique within the class of coordinate transformations and nonsingular input transformations. The stabilizing compensators S are, however, not unique; there are various classes of stabilizing compensators that can be used in a robust servomechanism controller. One of them is the *complementary controller* where the stabilizing compensator is given by

$$S: \quad \xi = K_0 x, \tag{2.15}$$

where x can either be the measured states of the system, or the estimates found using an observer (if not all states are measurable). In this case, the control u is given by state feedback found to stabilize the augmented system of the plant and the servocompensator:

$$\dot{\tilde{x}} = \tilde{A} \cdot \tilde{x} + \tilde{B} \cdot u,$$

$$\tilde{A} = \begin{bmatrix} A & 0 \\ -B_c C & A_c \end{bmatrix}, \qquad \tilde{B} = \begin{bmatrix} B \\ -B_c D \end{bmatrix}, \qquad \tilde{x} = \begin{bmatrix} x \\ \eta \end{bmatrix}, \qquad (2.16)$$

$$u = K_0 x + K_1 \eta = [K_0 \quad K_1 \tilde{x}] := K \cdot \tilde{x}.$$

A *perfect controller* for the robust servomechanism problem can be constructed using the *Cheap Control* method, where the gains $K = [\ K_0 \ K_1\ ]$ are found by minimizing the following performance index:

$$J_\varepsilon = \int\limits_0^\infty \left(z'z + \varepsilon \cdot u'u \right) d\tau, \quad z = \tilde{C} \cdot \tilde{x}, \qquad (2.17)$$

where $\varepsilon > 0$ is an arbitrarily small scalar. The optimal control gain $K = K_{opt}$, which minimizes this performance index, is given by $K_{opt} = -1/\varepsilon \tilde{B}' P_{opt}$, where P_{opt} is the unique positive semidefinite solution to the algebraic Riccati equation:

$$\tilde{A}' P_{opt} + P_{opt} \tilde{A} + \tilde{C}' \tilde{C} - \frac{1}{\varepsilon} P_{opt} \tilde{B} \tilde{B}' P_{opt} = 0. \qquad (2.18)$$

2.2.2 Discrete-Time Sliding Mode Control

Discrete time sliding mode controller is an approach in the sliding mode control theory, which is suitable for digital implementation since it does not exhibit the chattering phenomena due to direct digital implementation of continuous-time sliding mode control [3].

Consider a continuous linear time invariant system

$$\begin{aligned} \dot{x}(t) &= Ax(t) + Bu(t) + Ed(t), \\ y(t) &= Cx(t), \\ e(t) &= y(t) - y_{ref}(t). \end{aligned} \qquad (2.19)$$

with state vector $x(t) \in R^n$, control $u(t) \in R^m$, output to be regulated $y(t)$, reference input $y_{ref}(t)$, and disturbance $d(t)$. System (2.19) can be transformed to discrete time system with sampling time T_s to yield

$$\begin{aligned} x(k+1) &= A^* x(k) + B^* u(k) + E^* d(k), \\ y(k) &= Cx(k), \\ e(k) &= y(k) - y_{ref}(k), \end{aligned} \qquad (2.20)$$

where

$$A^* = \exp(A \cdot T_S), \qquad B^* = \int_0^{T_S} e^{A \cdot (T_S - \tau)B} \, d\tau,$$

$$E^* = \int_0^{T_S} e^{A \cdot (T_S - \tau)E} \, d\tau.$$

It is desired to control the output $y(k)$ to follow the reference $y_{ref}(k)$. For this purpose we can choose a sliding mode manifold in the form of $s(k) = Cx(k) - y_{ref}(k)$ such that when discrete-time sliding mode exists we have $y(k) \rightarrow y_{ref}(k)$. Discrete-time sliding mode exists if the control input $u(k)$ is designed as the solution of

$$s(k + 1) = CA^*x(k) + CB^*u(k) + y_{ref}(k + 1) = 0 \qquad (2.21)$$

The control law that satisfy (13) is called "equivalent control" and is given by

$$u_{eq}(k) = - \left(CB^* \right)^{-1} \left(CA^*x(k) - y_{ref}(k + 1) \right). \qquad (2.22)$$

If the control is limited to a value u_0 (i.e., $\|u(k)\| \leq u_0$) then the following modified control law can be applied:

$$u(k) = \begin{cases} \vec{u}_{eq}(k) & \text{for } \left\| \vec{u}_{eq}(k) \right\| \leq u_0, \\[2ex] u_0 \dfrac{\vec{u}_{eq}(k)}{\left\| \vec{u}_{eq}(k) \right\|} & \text{for } \left\| \vec{u}_{eq}(k) \right\| > u_0. \end{cases} \qquad (2.23)$$

With control law (2.16), discrete-time sliding mode exists after a finite number of steps.

2.3 CONTROL SYSTEM DEVELOPMENT

To achieve fast current limiting capability for the inverter, the control strategy uses a two-loop control structure: an inner inverter currents loop and an outer load voltages loop as shown in Fig. 2.5. The outer loop regulates the load voltages ($\bar{V}load_{qd}$) to follow 50/60 Hz balanced three-phase voltages references ($\vec{V}ref_{qd}$) and generates the inverter currents commands ($\vec{I}cmd_{qd}$), which are limited. The inner loop in turn generates the PWM voltage commands to regulate the inverter currents to follow the inverter current command. A standard voltage space vector algorithm is then used to realize these PWM command voltages. Notice that the 0-components of the load voltages are not regulated by the control, since it is uncontrollable.

Figure 2.6 shows the timing diagram of the PWM gating signals generation in relation with the A/D sampling time of the DSP. It can be seen that there is a one-half

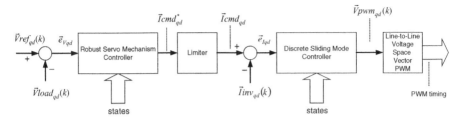

FIGURE 2.5   Overall control system.

PWM period delay between the time the signals are sampled by the A/D and the time the PWM control action is applied.

The next two subsections summarize the development of the two control loops, with the inner current loop using the discrete sliding mode controller and the voltage control loop using the robust servomechanism principles.

2.3.1   Discrete-Time Sliding Mode Current Controller

For designing the discrete-time sliding mode current controller, consider the inverter and filter subsystem with no transformer and load dynamics:

$$\frac{d\vec{V}inv_{qd}}{dt} = \frac{1}{3 \cdot C_{inv}}\vec{I}inv_{qd} - \frac{1}{3 \cdot C_{inv}}Tri_{qd0} \cdot \vec{I}snd_{qd0}, \qquad (2.24a)$$

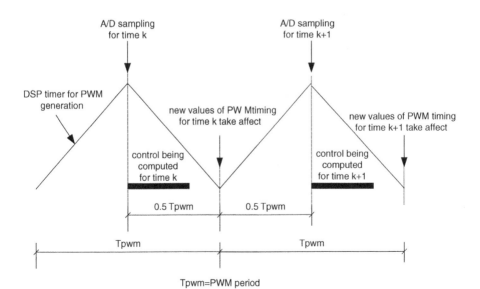

FIGURE 2.6   Control timing diagram.

$$\frac{d\vec{I}inv_{qd}}{dt} = \frac{1}{L_{inv}}\vec{V}pwm_{qd} - \frac{1}{L_{inv}}\vec{V}inv_{qd}. \tag{2.24b}$$

Assuming the secondary transformer currents $\vec{I}snd_{qd0}$ as disturbances, this subsystem can be written in state space form as

$$\dot{\vec{x}}_1 = A_1\vec{x}_1 + B_1\vec{u} + E_1\vec{d}_1,$$

$$\vec{A}_1 = \begin{bmatrix} \vec{0}_{2x2} & (3 \cdot C_{inv})^{-1} \cdot \vec{I}_{2x2} \\ -(L_{inv})^{-1} \cdot \vec{I}_{2x2} & \vec{0}_{2x2} \end{bmatrix},$$

$$\vec{B}_1 = \begin{bmatrix} \vec{0}_{2x2} \\ (L_{inv})^{-1} \cdot \vec{I}_{2x2} \end{bmatrix}, \quad \vec{E}_1 = \begin{bmatrix} -(3 \cdot C_{inv})^{-1} \cdot Tri_{qd0} \\ \vec{0}_{2x3} \end{bmatrix}, \tag{2.25}$$

where the states are $\vec{x}_1 = [\vec{V}inv_{qd}, \vec{I}inv_{qd}]$, the inputs $\vec{u} = \vec{V}pwm_{qd}$ and disturbances $\vec{d}_1 = \vec{I}snd_{qd}$.

The discrete form of Eqs. (2.25) can be calculated as

$$\vec{x}_1(k+1) = A_1^*\vec{x}_1(k) + B_1^*\vec{u}(k) + E_1^*\vec{d}_1(k),$$

where

$$A_1^* = \exp(A \cdot T_S), \quad B_1^* = \int_0^{T_S} e^{A_1 \cdot T_S} B_1 \, d\tau,$$
$$E_1^* = \int_0^{T_S} e^{A_1 \cdot T_S} E_1 \, d\tau$$

and T_S is the A/D sampling time, which in this case is equal to the PWM period T_{pwm}.

To force the inverter currents to follow their commands, the sliding mode surface is chosen as $\vec{s}(k) = C_1 \cdot \vec{x}_1(k) - \vec{I}cmd(k)$, where $C_1 \cdot \vec{x}_1(k) = \vec{I}inv_{qd}(k)$, so that when discrete sliding mode occurs, we have $\vec{s}(k) = 0$ or $\vec{I}inv(k) = \vec{I}cmd(k)$. The existence of the discrete sliding mode can be guaranteed if the control is given:

$$u(k) = \begin{cases} \vec{u}_{eq}(k) & \text{for } \|\vec{u}_{eq}(k)\| \le u_0, \\ u_0\dfrac{\vec{u}_{eq}(k)}{\|\vec{u}_{eq}(k)\|} & \text{for } \|\vec{u}_{eq}(k)\| > u_0, \end{cases} \tag{2.26}$$

where the equivalent control input $\vec{u}_{eq}(k)$ is calculated from

$$\vec{u}_{eq}(k) = (C_1 B_1^*)^{-1} \left(\vec{I}cmd_{qd} - C_1 A_1^*\vec{x}_1(k) - C_1 E_1^*\vec{d}_1(k) \right) \tag{2.27}$$

and u_0 denotes the maximum value of the PWM voltage command realizable by the space vector algorithm.

Note that the secondary transformer currents are needed for the control, but these currents are not measured in the system (see Fig. 3.2). A linear Luenberger observer can be easily designed to estimate these currents for control purposes. However, in most practical cases we can approximate these currents with the load currents (i.e., $\vec{I}snd_{qd} \approx \vec{I}load_{qd}$) since the currents through the output capacitor filters are small. According to the authors' experience, the effect of using this approximation is unnoticeable in the control performance.

Due to the computation delay of the DSP, the control action given by Eq. (2.27) will result in undesirable overshoots during transients. This effect can be minimized, however, if the states $\vec{x}_1(k)$ and disturbances $\vec{d}_1(k)$ are replaced with their first-order one-half step ahead predicted values given by

$$
\begin{aligned}
\vec{x}_1^p(k) &= 1.5 \cdot \vec{x}_1(k) - 0.5 \cdot \vec{x}_1(k-1), \\
\vec{d}_1^p(k) &= 1.5 \cdot \vec{d}_1(k) - 0.5 \cdot \vec{d}_1(k-1).
\end{aligned}
\tag{2.28}
$$

The equivalent control input $\vec{u}_{eq}(k)$ then becomes

$$
\vec{u}_{eq}(k) = \left(C_1 B_1^*\right)^{-1} \left(\vec{I}cmd_{qd} - C_1 A_1^* \vec{x}_1^p(k) - C_1 E_1^* \vec{d}_1^p(k)\right).
\tag{2.29}
$$

2.3.2 Voltage Controller Design Using Discrete Perfect RSP

The voltage control loop designed in this chapter is based on the discrete form of the technique developed in Section 2.3.1. To design the load voltages controller, let's first consider the entire plant system with the-0 components of the voltages and currents omitted as given in Eq. (30). As explained in Section 2.1, these 0-components are completely decoupled and uncontrollable from the inputs, and therefore they are not useful to be included in the design. In the system (2.30), an input delay of one-half the PWM period ($0.5T_{pwm}$) has been explicitly included to account for the computation delay of the DSP.

$$
\dot{x}_p(t) = A_p \vec{x}_p(t) + B_p \vec{u}(t - 0.5T_{pwm}),
$$

$$
\vec{A}_p =
\begin{bmatrix}
\vec{0}_{2x2} & (3 \cdot C_{inv})^{-1} \cdot \vec{I}_{2x2} & \vec{0}_{2x2} & -(3 \cdot C_{inv})^{-1} \cdot \hat{Tri}_{qd} \\
-(L_{inv})^{-1} \cdot \vec{I}_{2x2} & \vec{0}_{2x2} & \vec{0}_{2x2} & \vec{0}_{2x2} \\
\vec{0}_{2x2} & \vec{0}_{2x2} & \vec{0}_{2x2} & (C_{load})^{-1} \cdot \vec{I}_{2x2} \\
(L_{inv})^{-1} \cdot \hat{Trv}_{qd} & \vec{0}_{2x2} & -(L_{inv})^{-1} \cdot \vec{I}_{2x2} & -R_{trans}(L_{trans})^{-1} \cdot \vec{I}_{2x2}
\end{bmatrix},
$$

$$
\vec{B}_p =
\begin{bmatrix}
\vec{0}_{2x2} \\
(L_{inv})^{-1} \cdot \vec{I}_{2x2} \\
\vec{0}_{2x2} \\
\vec{0}_{2x2}
\end{bmatrix}, \quad
\hat{Tri}_{qd} = tr \cdot \frac{3}{2}
\begin{bmatrix}
1 & \sqrt{3} \\
-\sqrt{3} & 1
\end{bmatrix}, \quad
\hat{Trv}_{qd} = tr \cdot \frac{1}{2}
\begin{bmatrix}
1 & -\sqrt{3} \\
\sqrt{3} & 1
\end{bmatrix}.
\tag{2.30}
$$

The states variables for the system (2.30) are chosen as $\vec{x}_p = [\vec{V}inv_{qd}, \vec{I}inv_{qd}, \vec{V}load_{qd}, \vec{I}snd_{qd}]$ with the inputs as $u = \vec{V}pwm_{qd}$. System (2.30) can be transformed to a discrete-time system with sampling time $T_s = T_{pwm}$ to yield

$$\vec{x}_p(k+1) = \Phi \cdot \vec{x}_p(k) + \Gamma_1 \cdot \vec{u}(k-1) + \Gamma_2 \cdot \vec{u}(k), \tag{2.31}$$

where

$$\Phi = e^{A_p T_s}, \quad \Gamma_1 = \int_{0.5T_s}^{T_s} e^{A_p \tau} B_p \, d\tau, \quad \Gamma_2 = \int_0^{0.5T_s} e^{A_p \tau} B_p \, d\tau.$$

Discrete time system (2.31) can be written in a standard discrete-time state-space equations by adding the extra states: $\vec{x}_a(k) = \vec{u}(k-1) = \vec{V}pwm_{qd}(k-1)$ to yield

$$\begin{bmatrix} \vec{x}_p(k+1) \\ \vec{x}_a(k+1) \end{bmatrix} = \begin{bmatrix} \Phi & \Gamma_1 \\ \vec{0}_{2x2} & \vec{0}_{2x2} \end{bmatrix} \cdot \begin{bmatrix} \vec{x}_p(k) \\ \vec{x}_a(k) \end{bmatrix} + \begin{bmatrix} \Gamma_2 \\ \vec{I}_{2x2} \end{bmatrix} \cdot \vec{u}(k) \tag{2.32}$$

so that the system can be written as

$$\vec{x}_p^*(k+1) = A_p^* \vec{x}_p^*(k) + B_p^* \vec{u}(k), \tag{2.33}$$

where

$$\vec{x}_p^*(k) = \begin{bmatrix} \vec{x}_p(k) \\ \vec{x}_a(k) \end{bmatrix}, \quad A_p^* = \begin{bmatrix} \Phi & \Gamma_1 \\ \vec{0}_{2x2} & \vec{0}_{2x2} \end{bmatrix}, \quad B_p^* = \begin{bmatrix} \Gamma_2 \\ \vec{I}_{2x2} \end{bmatrix}.$$

To design the *voltage controller*, we need to consider the true plant (2.33) and the discrete time sliding mode current controller as the equivalent "plant" as seen by the outer voltage loop. Using Eqs. (2.27) and (2.33) the augmented true plant and discrete sliding mode current controller can be found as in (2.34):

$$\vec{x}_p^*(k+1) = A_d \vec{x}_p^*(k) + B_d \vec{u}_1(k) \tag{2.34}$$

with $\vec{u}_1(k) = \vec{I}cmd^*_{qd}(k)$, and

$$A_d = A^*_p - B^*_p \left(C_1 B^*_1\right)^{-1} \left(B^*_1 C_{11} + E^*_1 C_{12}\right),$$

$$B_d = B^*_p \left(C_1 B^*_1\right)^{-1},$$

$$C_{11} = \begin{bmatrix} \vec{I}_{2x2} & \vec{0}_{2x2} & \vec{0}_{2x2} & \vec{0}_{2x2} & \vec{0}_{2x2} \\ \vec{0}_{2x2} & \vec{I}_{2x2} & \vec{0}_{2x2} & \vec{0}_{2x2} & \vec{0}_{2x2} \end{bmatrix},$$

$$C_{12} = \begin{bmatrix} \vec{0}_{2x2} & \vec{0}_{2x2} & \vec{0}_{2x2} & \vec{I}_{2x2} & \vec{0}_{2x2} \end{bmatrix}.$$

Note that the augmented system given in Eq. (2.34) was found, assuming that the approximation $\vec{I}snd_{qd} \approx \vec{I}load_{qd}$ has been used.

Now, assume that $\omega_i = 2\pi f_i$, $i = 1, 2, \ldots n$, are frequencies of the reference voltages and harmonics to be eliminated. For a 60-Hz UPS system with desire to eliminate fifth and seventh harmonics, for example, we use $\omega_1 = 2\pi \cdot 60$, $\omega_2 = 2\pi \cdot 5 \cdot 60$, and $\omega_3 = 2\pi \cdot 7 \cdot 60$. We can then choose the servocompensator to be of the form (2.35)

$$
\begin{aligned}
\dot{\vec{\eta}} &= A_c \vec{\eta} + B_c e_{Vqd}, \\
\vec{e}_{Vqd} &= \vec{V}ref_{qd} - \vec{V}load_{qd},
\end{aligned}
\tag{2.35}
$$

where

$$\vec{\eta} = \begin{bmatrix} \vec{\eta}_1, & \vec{\eta}_2, \ldots, & \vec{\eta}_n \end{bmatrix}^T, \quad \vec{\eta}_i \in R^4, \quad i = 1, 2, \ldots, n,$$

$$A_c = \text{block diag } [Ac_1, \ Ac_2, \ldots, \ Ac_n],$$

$$B_c = [Bc_1, \ Bc_2, \ldots, \ Bc_n]^T$$

with

$$Ac_i = \begin{bmatrix} \vec{0}_{2x2} & \vec{I}_{2x2} \\ -\omega_i^2 \vec{I}_{2x2} & \vec{0}_{2x2} \end{bmatrix}, \quad i = 1, 2, \ldots, n,$$

$$Bc_i = \left(\vec{0}_{2x2} \quad \vec{I}_{2x2}\right)^T, \quad i = 1, 2, \ldots n.$$

Note that each of the blocks $\dot{\vec{\eta}}_i = Ac_1 \vec{\eta}_i + Bc_i \vec{e}_{Vqd}$ represents a state space implementation of the continuous transfer function: $1/\left(s^2 + \omega_i^2\right)$ for each of the qd-axis voltages errors.

The servocompensator (2.35) can be transformed to a discrete time system to yield

$$\vec{\eta}(k + 1) = A^*_c \vec{\eta}(k) + B^*_c \vec{e}_{Vqd}(k), \quad \vec{e}_{Vqd}(k) = \vec{V}ref_{qd}(k) - \vec{V}load_{qd}(k), \tag{2.36}$$

where

$$A_c^* = \exp(A_c \cdot T_S), \quad B_c^* = \int_0^{T_S} e^{A_1 \cdot (T_S - \tau)} B_c \, d\tau.$$

Now that we have determined the "plant" and the servocompensator, the control input for the *perfect robust servomechanism controller* is given by

$$\vec{u}_1(k) = \vec{I}cmd_{qd}^*(k) = K_0 x_p^*(k) + K_1 \eta(k) \tag{2.37}$$

where the gains $K = [K_0 \ K_1]$ are found by minimizing the discrete performance index:

$$J_\varepsilon = \sum_{k=0}^{\infty} \left(z(k)' z(k) + \varepsilon \cdot u(k)' u(k) \right),$$

$$z = \begin{bmatrix} x_p^* \\ \eta \end{bmatrix} \tag{2.38}$$

for the augmented "equivalent plant" (2.34) and the servocompensator (2.36):

$$\begin{bmatrix} \vec{x}_p^*(k+1) \\ \eta(k+1) \end{bmatrix} = \begin{bmatrix} A_d & 0 \\ -B_c^* C & A_c^* \end{bmatrix} \begin{bmatrix} \vec{x}_p^*(k) \\ \eta(k) \end{bmatrix} + \begin{bmatrix} B_d \\ -B_c^* D \end{bmatrix} u_1(K), \tag{2.39}$$

where $\varepsilon > 0$ is an arbitrarily small scalar.

2.3.3 Control of Current Limit and Saturation

The current command $\vec{I}cmd_{qd}^*(k)$ generated by the robust servomechanism voltage controller above is limited in magnitude as in Eq. (2.40) to yield the current command $\vec{I}cmd_{qd}(k)$, which will be implemented by the inner loop current controller:

$$\vec{I}cmd_{qd}(k) = \begin{cases} \vec{I}cmd_{qd}^*(k) & \text{if } \left| \vec{I}cmd_{qd}^*(k) \right| \leq I_{max}, \\[2mm] \dfrac{\vec{I}cmd_{qd}^*(k)}{\left| \vec{I}cmd_{qd}^*(k) \right|} I_{max} & \text{if } \left| \vec{I}cmd_{qd}^*(k) \right| > I_{max}, \end{cases} \tag{2.40}$$

Where I_{max} represents the maximum allowable magnitude of the inverter currents. Equation (2.40) limits the magnitude of the current commands but maintains their vector directions in the qd-space.

The states $\vec{\eta}_i$ of the servocompensator can be seen as sine wave signal generators that get excited by the harmonic contents of the error signals at frequency ω_i. When the control inputs of the robust servomechanism voltage controller saturate—that is, $|\vec{I}cmd_{qd}^{*}\ k| > I_{max}$—the servocompensator states will grow in magnitude due to the break in the control loop. This problem is similar to the integrator windup problem that occurs in an integral-type controller. To prevent this, the servocompensator in (2.36) can be modified as follows:

$$\vec{\eta}(k+1) = A_{c}^{*}\vec{\eta}(k) + B_{c}^{*}\vec{e}_{1}(k),$$

$$\vec{e}_{1}(k) = \begin{cases} \vec{e}_{Vqd}(k) & \text{if } |\vec{I}cmd_{qd}^{*}| \leq I_{max}, \\ \\ 0 & \text{if } |\vec{I}cmd_{qd}^{*}| > I_{max}. \end{cases} \qquad (2.41)$$

Using (2.41), during the current limit saturation, the servocompensator states will continue to oscillate at the harmonic frequency with constant magnitude. The resulting robust servomechanism controller structure is shown in Fig. 2.7.

2.4 STEP-BY-STEP CONTROL FLOW EXPLANATIONS

1. At each control sampling time k the following voltages and currents signals are sampled.
   - The line-to-neutral load voltages (at points xyz-n in Fig. 2.8) are denoted as $Vload_{an}$, $Vload_{bn}$, and $Vload_{cn}$.
   - The load phase currents are denoted as $Iload_{a}$, $Iload_{b}$, and $Iload_{c}$.
   - The line-to-line inverter filter capacitor voltages (at points uvw in Fig. 2.8) are denoted as $Vinv_{ab}$, $Vinv_{bc}$, and $Vinv_{ca}$.
   - The inverter phase currents are denoted as $Iinv_{a}$, $Iinv_{b}$, and $Iinv_{c}$.

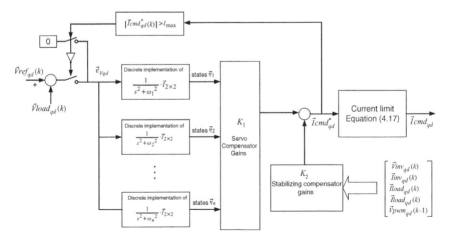

FIGURE 2.7 Output voltages controller using robust servomechanism controller.

2. Form the following voltages and currents vectors in *abc* reference frame:

$$\vec{V}inv_{abc} = \begin{bmatrix} Vinv_{ab} & Vinv_{bc} & Vinv_{ca} \end{bmatrix}^T \quad \text{abc inverter voltages vector,}$$

$$\vec{V}load_{abc} = \begin{bmatrix} Vload_a & Vload_b & Vload_c \end{bmatrix}^T \quad \text{abc load voltages vector,}$$

$$\vec{I}load_{abc} = \begin{bmatrix} Vload_a & Iload_b & Iload_c \end{bmatrix}^T \quad \text{abc load currents vector,}$$

$$\vec{I}inv_{abc} = \begin{bmatrix} Vinv_{ab} & Iinv_{bc} & Iinv_{ca} \end{bmatrix}^T \quad \text{abc inverter currents vector.}$$

$$= \begin{bmatrix} Iinv_a - Iinv_b & Iinv_b - Iinv_c & Iinv_c - Iinv_a \end{bmatrix}^T$$

Note that the inverter currents vectors are represented as line-to-line quantities. This is done to match the chosen state space equations and state variables used to derive the control. Inverter phase currents can also be used here with proper modification to the state-space equations used.

3. The sampled voltages and currents are then transformed to their DQ0 stationary reference frame using the transformation:

$$\vec{f}_{qd0} = K_S \cdot \vec{f}_{abc},$$

with

$$K_S = \frac{2}{3} \begin{bmatrix} 1 & -0.5 & -0.5 \\ 0 & -\sqrt{3}/2 & \sqrt{3}/2 \\ 0.5 & 0.5 & 0.5 \end{bmatrix},$$

$$f_{qd0} = \begin{bmatrix} f_q, & f_d, & f_0 \end{bmatrix}^T, [f_{abc} = f_a, f_b, f_c]^T.$$

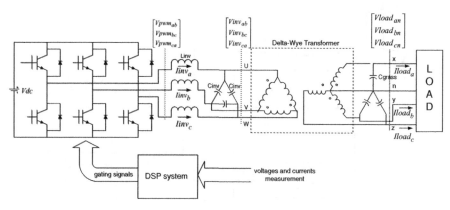

FIGURE 2.8 Power converter system (see Fig. 2.1).

The following voltages and currents variables in DQ stationary reference frame are obtained:

$$\vec{V}inv_{qd0} = \vec{K}_S \cdot \vec{V}inv_{abc} = \begin{bmatrix} Vinv_q & Vinv_d & Vinv_0 \end{bmatrix}^T$$

$$\vec{I}inv_{qd0} = \vec{K}_S \cdot \vec{I}inv_{abc} = \begin{bmatrix} Vinv_q & Iinv_d & Iinv_0 \end{bmatrix}^T$$

$$\vec{V}load_{qd0} = \vec{K}_S \cdot \vec{V}load_{abc} = \begin{bmatrix} Vload_q & Vload_d & Vload_0 \end{bmatrix}^T$$

$$\vec{I}load_{qd0} = \vec{K}_S \cdot \vec{I}load_{abc} = \begin{bmatrix} Vload_q & Iload_d & Iload_0 \end{bmatrix}^T$$

The 0-component of the inverter voltages and currents will always be zero and can be discarded due to the three-wire system of the inverter. The 0-components of the load voltages and currents can be nonzero; however, these components are noncontrollable and have no use for control. Therefore only the DQ components need to be used and calculated:

$$\vec{V}inv_{qd} = \begin{bmatrix} Vinv_q & Vinv_d \end{bmatrix}^T \quad \vec{V}load_{qd} = \begin{bmatrix} Vload_q & Vload_d \end{bmatrix}^T,$$

$$\vec{I}inv_{qd} = \begin{bmatrix} Iinv_q & Iinv_d \end{bmatrix}^T \quad \vec{I}load_{qd} = \begin{bmatrix} Iload_q & Iload_d \end{bmatrix}^T.$$

Note that the DQ0 transformation used here is a standard transformation commonly used in three-phase power applications. The transformation puts the three-phase variables in the system into an orthogonal coordinate system. The use of an orthogonal coordinate system allows the identification of state variables and creation of the state-space equations for control purposes.

4. Generate the sine waves references at discrete time k for the load voltages in the *abc* reference frame and perform the DQ transformation on them to obtain the references in DQ.

$$Vref_{an} = V_{mA} \cdot \sin(2 \cdot \pi \cdot f_{ref} \cdot k \cdot T_S),$$

$$Vref_{bn} = V_{mB} \cdot \sin(2 \cdot \pi \cdot f_{ref} \cdot k \cdot T_S - 2\pi/3),$$

$$Vref_{cn} = V_{mC} \cdot \sin(2 \cdot \pi \cdot f_{ref} \cdot k \cdot T_S + 2\pi/3),$$

where

V_{mA}	: voltage magnitude of reference AN
V_{mB}	: voltage magnitude of reference BN
V_{mC}	: voltage magnitude of reference CN
T_S	: sampling period
f_{ref}	: fundamental reference frequency

Transform to DQ reference frame:

$$\vec{V}ref_{abc} = \begin{bmatrix} Vref_{an} & Vref_{bn} & Vref_{cn} \end{bmatrix}^T$$

$$\vec{V}ref_{qd0} = \vec{K}_S \cdot \vec{V}ref_{abc} = \begin{bmatrix} Vref_q & Vref_d & Vref_{cn} \end{bmatrix}^T$$

Again only the DQ components are needed for control:

$$\vec{V}ref_{qd} = \begin{bmatrix} Vref_q & Vref_d \end{bmatrix}$$

5. Compute the load voltages errors to be used by the voltage servo compensator:

$$e_q = Vref_q - Vload_q,$$

$$e_d = Vref_d - Vload_d.$$

6. Calculate the next discrete states (time $k + 1$) of the servocompensator for each of the harmonic frequencies to be eliminated and the fundamental frequency and for each Q- and D-axis voltages errors. The following diagram shows the case when the fifth and seventh are to be eliminated.

Each harmonic servocompensator block in Fig. 2.9 is the discrete state space implementation of the analog transfer function:

$$T(s) = \frac{1}{s^2 + \omega_n^2},$$

where $\omega_i = 2\pi \cdot i \cdot f$, f is the fundamental frequency and $i \cdot f$ is the harmonic frequency (i.e in the case of Fig. 2.9 $i = 1, 5,$ and 7).

Specifically, the following continuous state space implementation of $T(s)$ is used for each harmonic servocompensator block:

$$\dot{\vec{\eta}}_i = \begin{pmatrix} 0 & 1 \\ -\omega_i^2 & 0 \end{pmatrix} \vec{\eta}_i + \begin{pmatrix} 0 \\ 1 \end{pmatrix} e$$

which is then discretized to obtain the a's and b's coefficients shown in Fig. 2.9. Standard discretization method like ZOH (zero-order hold) method is used in this case.

Each harmonic servocompensator block in Fig. 2.9 ensures the elimination of the corresponding harmonic contents of the errors signals. This condition is guaranteed by the *internal model principle* that states that if frequency modes (poles) of the references and the disturbances to be eliminated are included in the control loop, then the steady-state error will not contain these frequency contents. Each of these harmonic compensators can be viewed as signals generators that vary at the harmonic

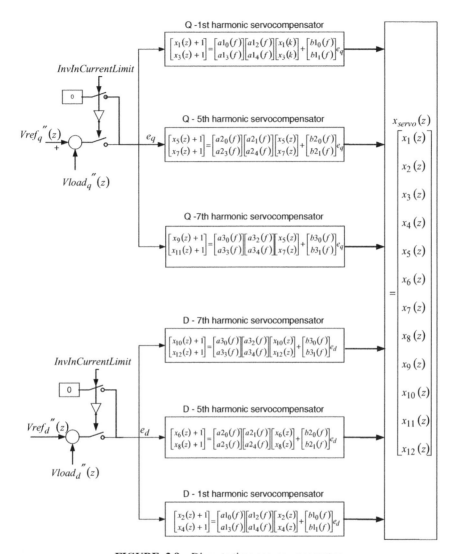

FIGURE 2.9   Discrete time servocompensator.

frequency and reacts only to the existence of the corresponding harmonic in the errors signals. It provides adjustment to the inverter current commands generated by the voltage control loop to force the corresponding harmonic content of the errors to zero.

Multiply each state of the servocompensator states $(x_{servo}(k))$ with the corresponding servocompensator gain and get the overall sum for each axis as shown in Fig. 2.10. Multiply the plant states (consisting of the inverter voltages, inverter currents, load voltages, load currents, and the previous PWM voltage commands) with the stabilizing compensator gains and get the summation for each axis. The sum of

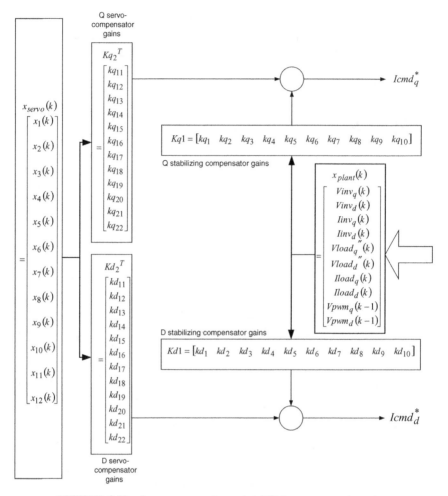

FIGURE 2.10 Servocompensator and stabilizing compensator gains.

the servocompensator and the stabilizing compensator for each axis constitutes the inverter current commands generated by the voltage control loop, that is, $Icmd_q$ and $Icmd_d$.

The servocompensator and the stabilizing compensator gains are designed to ensure the stability of the overall control loop—that is, the combined voltage control loop and the current control loop. The *perfect RSP theory* provides a way of calculating these gains using an *OPTIMAL Control Technique*. Using this technique, a cost function consisting of the weighted sum of squares of the servocompensator states, the plant states, and the control inputs (inverter current commands) is created. An optimization routine is then used to seek the gains that minimize this cost function. The weighted sum of squares of the servocompensator states in this case can be viewed as the energy in the errors signals, while those of the plant states and the

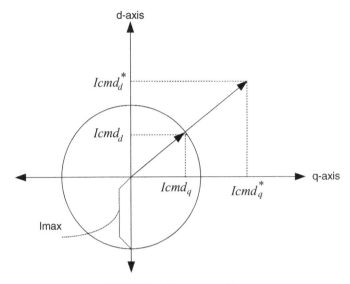

FIGURE 2.11 Current limit.

control inputs can be viewed as the fuel energy needed. Different weighting used can be used to obtain different behavior of the control system.

Perform the current limiting function. The purpose of this current limiting function is to limit the inverter currents commands generated by the voltage controller. At each sampling time the inverter currents commands generated by the voltage controller will be implemented by the inner current loop which will compute the necessary PWM command. The current limiting is accomplished by limiting the magnitude of the inverter current commands vector and maintaining its original direction as shown in Fig. 2.11. The current limit appears as a circle in the DQ stationary reference frame. If the command vector generated by the voltage controller goes outside of this circle, its magnitude is limited to be within the circle with its direction unaltered. Limiting the inverter currents in this way ensures that at each time the magnitudes of the three-phase inverter currents in the *abc* reference frame remain within the limit value but still provides the phase relations required by the voltage control loop.

Mathematically, this limiting function can written as

$$
\vec{I}cmd_{qd}(k) = \begin{cases} \vec{I}cmd_{qd}^{*}(k) & \text{if } \left| \vec{I}cmd_{qd}^{*}(k) \right| \leq I_{\max}, \\[2ex] \dfrac{\vec{I}cmd_{qd}^{*}(k)}{\left| \vec{I}cmd_{qd}^{*}(k) \right|} I_{\max} & \text{if } \left| \vec{I}cmd_{qd}^{*}(k) \right| > I_{\max}, \end{cases}
$$

where absolute value indicates square root of sum of the squares.

Compute the errors between the limited inverter current commands and the actual inverter commands:

$$e_{Iq} = Icmd_q - Iinv_q,$$

$$e_{Id} = Icmd_d - Iinv_d.$$

The discrete sliding mode current controller will force these errors to zero at the next sampling time $(k + 1)$ by computing the necessary PWM voltage command.

7. Compute the one-half step ahead prediction of the inverter voltages and currents and the load currents. These predicted values of the voltages and currents are necessary to account for the delay in the application of the PWM voltage due to DSP computation time.

$$\vec{x}_1^P(k) = 1.5 \cdot \vec{x}_1(k) - 0.5 \cdot \vec{x}_1(k - 1),$$

$$\vec{d}_1^P(k) = 1.5 \cdot \vec{d}_1(k) - 0.5 \cdot \vec{d}_1(k - 1).$$

8. Compute the necessary PWM voltages to implement the inverter current command using the discrete sliding mode current controller. The discrete sliding mode current controller ensures that the at the next sampling time $(k + 1)$, the inverter current commands obtain the values of the command using the available DC bus voltage at the time.

The necessary PWM voltages commands are given by

$$\vec{V}pwm_{qd} = u(k) = \begin{cases} \vec{u}_{eq}(k) & \text{for } \|\vec{u}_{eq}(k)\| \leq u_0, \\ u_0 \dfrac{\vec{u}_{eq}(k)}{\|\vec{u}_{eq}(k)\|} & \text{for } \|\vec{u}_{eq}(k)\| > u_0, \end{cases}$$

$$\vec{u}_{eq}(k) = \left(C_1 B_1^*\right)^{-1} \left(\vec{I}cmd_{qd} - C_1 A_1^* \vec{x}_1^P(k) - C_1 E_1^* \vec{d}_1^P(k)\right).$$

Note that the PWM voltage commands are limited by the amount of available DC bus voltage at that time. This limit value is denoted as u_o in the above equations.

Finally, the PWM voltage commands are implemented into actual PWM voltages using a standard space vector PWM technique. The space vector PWM technique takes the PWM voltages commands $Vpwm_q$ and $Vpwm_d$ and computes the necessary timing to switch on and off the power switches.

2.5 MATLAB PROGRAM TO CALCULATE THE CONTROL GAINS

This section provides a step-by-step explanation on how to use a MATLAB program to calculate the control gains. The full listing of the MATLAB script can be found in the Appendix. The file is stored as 'gains.m'.

Step 1: Define circuit parameters and constants necessary for the control gains calculations.
Step 1a. Define circuit parameters:

```
Cinv=540e-6;                    % Inverter capacitor filter
Linv=298e-6;                    % Inverter inductor filter
Cload=90e-6;                    % Grass capacitor
Ltrans=0.03*208*208/80e3/wfun;  % 3% p.u transformer
                                  inductance
Rtrans=0.03*208*208/80e3;       % 3% p.u transformer
                                  resistance
tr=120/245;                     % Inverter To Output
                                  Turn ratio
Ilimit=3*80e3/245*sqrt(2);      % 300% inverter current
                                  limit
```

Step 1b. Define delta-wye transformer voltages and currents transfer matrices [see Eqs. 2.1–2.6]:

```
Tri_qd=3/2*tr*[1 sqrt(3);
    -sqrt(3) 1 ];
Trv_qd=1/2*tr*[1 -sqrt(3);
      sqrt(3) 1 ];
```

Step 1c. Define the fundamental output frequency and the harmonics;

```
ffun=60;
wfun=2*pi*ffun;
w1=wfun;
w2=2*wfun;
w3=3*wfun;
w5=5*wfun;
w4=4*wfun;
w7=7*wfun;
T1=1/ffun;
```

Step 1d. Define control sampling time:

```
Tsamp=1/60/51;    %6536/20e6;
```

Step 2: Design the control gains for the sliding mode current controller.
Step 2a. Define the plant for the current controller [see EQ. 2.8–2.12]:

```
A=[zeros(2,2)          eye(2,2)/(3*Cinv)  ;
-1/Linv*eye(2)        -0.01/Linv*eye(2)];
B=[zeros(2,2);
1/Linv*eye(2)];
```

```
E=[-Tri_qd/(3*Cinv);
zeros(2,2)];
F=[zeros(2,2);
-eye(2)];
C=[zeros(2,2) eye(2)];
D=zeros(2,2);
```

Step 2b. Discretize the plant for the current controller:

```
sysc=ss(A,B,C,zeros(size(C,1),size(B,2)));
sysd=c2d(sysc,Tsamp,'zoh');
[Acurrd,Bcurrd,Ccurrd,Dcurrd]=ssdata(sysd);
```

Step 2c. Calculate the gains of the equivalent control:

```
CBinv=inv(Ccurrd*Bcurrd);
CA=Ccurrd*Acurrd;
CD=Ccurrd*F;
sysc=ss(A,E,C,zeros(size(C,1),size(B,2)));
sysd=c2d(sysc,Tsamp,'zoh');
[Acurrd1,Ecurrd,Ccurrd1,Dcurrd1]=ssdata(sysd);
CE=Ccurrd1*Ecurrd;
```

Step 3: Design the control gains for the servovoltage controller.
Step 3a. Define the true plant:

```
Ao=[zeros(2,2)   eye(2,2)/(3*Cinv) zeros(2,2) -Tri_qd/(3*Cinv);
  -eye(2,2)/Linv zeros(2,2)   zeros(2,2)   zeros(2,2);
  zeros(2,2)    zeros(2,2)    zeros(2,2)    eye(2,2)/Cload;
Trv_qd/Ltrans zeros(2,2) -eye(2,2)/Ltrans -Rtrans*eye(2,2)/Ltrans];
Bo=[zeros(2,2);
  eye(2,2)/Linv;
  zeros(2,2);
  zeros(2,2)];

Co=[zeros(2,2) zeros(2,2) eye(2) zeros(2,2)];
```

Step 3b. Define the analog servocompensator for each harmonic to be eliminated:

```
Ch0=zeros(2,2);

Ch1=[zeros(2,2)   eye(2);
-w1^2*eye(2) zeros(2,2)];

Ch5=[zeros(2,2)   eye(2);
-w5^2*eye(2) zeros(2,2)];
```

```
Ch7=[zeros(2,2)   eye(2);
-w7^2*eye(2) zeros(2,2)];
```

Step 3b. Form the servocompensator state matrices (see Eqs. 2.14–2.16]:

```
Ch_star=[Ch1   eros(size(Ch1))     zeros(size(Ch1));
  zeros(size(Ch1))    Ch5    zeros(size(Ch1));
  zeros(size(Ch1))    zeros(size(Ch1)) Ch7];

Bh_star=[zeros(2,2);
  eye(2,2);
  zeros(2,2);
  eye(2,2);
  zeros(2,2);
  eye(2,2)];
```

Step 3c. Discretize true plant:

```
sysc=ss(Ao,Bo,Co,zeros(size(C,1),size(B,2)));
sysd=c2d(sysc,Tsamp,'zoh');
[Aod,Bod,Cd,Dd]=ssdata(sysd);
```

Step 3d. Calculate equivalent combined plant+ DSM current controller:

```
C1=[eye(4)   zeros(4,4)];
C2=[zeros(2,4)   eye(2) zeros(2,2)];
Ad=Aod-Bod*(CBinv*CA*C1+CBinv*CE*C2);
Bd=Bod*CBinv;
```

Step 3e. Discetize the controller:

```
csysc=ss(Ch_star,Bh_star,eye(size(Ch_star,1)),
  zeros(size(Ch_star,1),size(Bh_star,2)));
[csysbc,T]=ssbal(csysc);
csysd=c2d(csysbc,Tsamp,'zoh');
[Acon_d,Bcon_d,Ccon_d,Dcon_d]=ssdata(csysd);
```

Step 3f. Form the augmented equivalent plant and the servocompensator:

```
Ad_big=[Ad    zeros(size(Ad,1),size(Acon_d,2));
    -Bcon_d*Cd    Acon_d];
Bd_big=[Bd;-Bcon_d*Dd];
```

Step 3g. Define the weighting matrices:

```
epsilon=1e-5;
Q2=eye(size(Acon_d,1));
```

```
state_W=0.2; % 80 kVA
Q1=state_W*eye(size(Ad,1));
Q=[Q1  zeros(size(Ad,1),size(Acon_d,2));
zeros(size(Acon_d,1),size(Ad,2))    5e5*Q2];
R=epsilon*eye(2);
```

Step 3h. Now performed the optimal calculations of the gains:

```
[Kd,S,E]=dlqr(Ad_big,Bd_big,Q,R);
Kd=-Kd;
```

2.6 SIMULATION USING SIMULINK AND MATLAB

Simulations of the control algorithm are done using Simulink MATLAB. This section describes the simulation assumptions and the Simulink block diagrams used.

2.6.1 Simulation Assumptions

The following assumptions are used in building the simulation:

1. It is assumed the PWM space vector and the power converter generate ideal voltages according to the commands given by the control. This assumption is valid due to the relatively high switching frequency of the PWM inverter.
2. The three-phase delta–wye transformer is modeled as three single-phase ideal transformers with series leakage resistances. This is the same model used in the control derivations.

2.6.2 Simulation Block Diagrams: The Main Simulink Block Diagrams

Figure 2.12 shows the main simulation diagram in Simulink. The diagram can be divided into four main parts:

Voltage References: This part generates the sine wave references for the voltages. It can provide generation of sine wave at fundamental and harmonics frequencies
Plant (Power Converter + Load): This part implements the power converter, filter, and the load.
Digital Controller: This part implements the voltage and current controller.
Display (Scopes and Meter): This part provides the waveforms displays and metering.

2.6.3 The Plant

The plant (i.e., the power converter, the filter, and the load) is simulated using a Simulink continuous state-space block. Parameters A, B, C, and D of the plant state space are initialized from the MATLAB workspace. The following discussion gives step-by-step instructions explaining how to initialize the *A*, *B*, *C*, and *D* parameters of the plant for different types of loads. Note that the *B* parameter remains the same for different types of loads available.

Step 1. Declare some constants and matrices.

```
Rfl=208*208/(80e3*0.8);         % Full load resistance
Rtrans=0.03*208*208/80e3;       % transformer series resistance
```

Step 2. Define transformer delta-wye transfer matrices.

```
Tri=[1 -2 1;
  1  1 -2;
  -2  1 1];

Trv=[0    0 -1;
  -1 0 0;
  0 -1 0];


Tri_qd0=3/2*tr*[1 sqrt(3) 0;
  -sqrt(3) 1 0];

Trv_qd0=1/2*tr*[1 -sqrt(3);
  sqrt(3) 1;
     0    0 ];
```

Step 3. Define ABC to DQ0 transformation matrix.

```
Ks=2/3*[cos(0) cos(0-2*pi/3) cos(0+2*pi/3);
  sin(0) sin(0-2*pi/3) sin(0+2*pi/3);
     1    1    1 ];

Ksinv=inv(Ks)   ;
```

Step 4. Define the B and D matrices which will be the same for all types of load.

```
Bsim=[zeros(2,2);
eye(2,2)/Linv;
  zeros(3,2);
  zeros(3,2)];
```

Step 4. Define the A and C matrices for different types of loads.
Step 4a. Define the A and C matrices for full load resistive.
Define the resistive load matrix for full load values:

```
Rload33=[ Rfl  0  0;
    0  Rfl  0;
     0  0  Rfl];
```

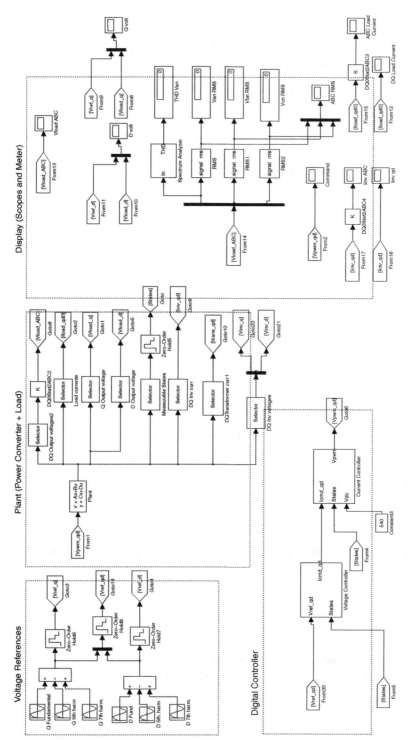

FIGURE 2.12 Main Simulink simulation diagrams.

Form A matrix for full load resistive:

```
Afl=[ zeros(2,2)  eye(2,2)/(3*Cinv)  zeros(2,3)  -Tri_qd0/(3*Cinv);
-eye(2,2)/Linv zeros(2,2)  zeros(2,3)  zeros(2,3);
zeros(3,2)  zeros(3,2)  -Ks*inv(Rload33)*Ksinv/Cload  -eye(3,3)/Cload;
Trv_qd0/Ltrans zeros(3,2)  -eye(3,3)/Ltrans  -Rtrans*eye(3,3)/Ltrans];
```

Form C matrix for full load resistive:

```
Cfl=[ [eye(10)];
   [zeros(3,4) Ks*inv(Rload33)*Ksinv zeros(3,3)]];
```

Step 4b. Define the A and C matrices for no load.
Define the resistive load matrix for no load:

```
Rload33=[100  0  0;
  0  100  0;
  0  0  100];
```

Form A matrix for no load resistive:

```
Anl=[zeros(2,2)  eye(2,2)/(3*Cinv)  zeros(2,3)  -Tri_qd0/(3*Cinv);
-eye(2,2)/Linv zeros(2,2)  zeros(2,3)  zeros(2,3);
zeros(3,2)  zeros(3,2)  -Ks*inv(Rload33)*Ksinv/Cload  eye(3,3)/Cload;
Trv_qd0/Ltrans zeros(3,2)  -eye(3,3)/Ltrans  -Rtrans*eye(3,3)/Ltrans];
```

Form C matrix for no load resistive:

```
Cnl=[ [eye(10)];
  [zeros(3,4) Ks*inv(Rload33)*Ksinv zeros(3,3)]];
```

Step 4c. Define the A and C matrices for single-phase 100% load.
Define the resistive load matrix for single-phase 100% load:

```
Rload33=[ Rfl    0    0;
  0    100    0;
  0    0    100];
```

Form A matrix for single phase 100% load:

```
Apl=[zeros(2,2)  eye(2,2)/(3*Cinv)  zeros(2,3)  -Tri_qd0/(3*Cinv);
-eye(2,2)/Linv zeros(2,2)  zeros(2,3)  zeros(2,3);
zeros(3,2)  zeros(3,2)  -Ks*inv(Rload33)*Ksinv/Cload  eye(3,3)/Cload;
Trv_qd0/Ltrans zeros(3,2)  -eye(3,3)/Ltrans  -Rtrans*eye(3,3)/Ltrans];
```

Form C matrix for single phase 100% load:

```
Cpl=[ [eye(10)];
   [zeros(3,4) Ks*inv(Rload33)*Ksinv zeros(3,3)]];
```

Step 4d. Define the A and C matrices for two-phase 100% resistive load.
Define the resistive load matrix for two-phase 100% load:

```
Rload33=[ Rfl    0    0;
          0    Rfl    0;
          0     0   100];
```

Form A matrix for two phase 100% load:

```
Ap2=[zeros(2,2)  eye(2,2)/(3*Cinv)  zeros(2,3)  -Tri_qd0/(3*Cinv);
-eye(2,2)/Linv zeros(2,2)  zeros(2,3)  zeros(2,3);
zeros(3,2)  zeros(3,2)  -Ks*inv(Rload33)*Ksinv/Cload  eye(3,3)/Cload;
Trv_qd0/Ltrans  zeros(3,2)  -eye(3,3)/Ltrans  -Rtrans*eye(3,3)/Ltrans];
```

Form A matrix for two phase 100% loaded:

```
Cp2=[ [eye(10)];
   [zeros(3,4) Ks*inv(Rload33)*Ksinv zeros(3,3)]];
```

Step 4e Define the A and C matrices for 500% resistive load.
Define the resistive load matrix for 500% load:

```
Rload33=[ sqrt(0.2*Rfl)    0    0;
          0    sqrt(0.2*Rfl)    0;
          0     0    sqrt(0.2*Rfl)];
```

Form A matrix for 500% load:

```
Aov=[  zeros(2,2)  eye(2,2)/(3*Cinv)  zeros(2,3)  -Tri_qd0/(3*Cinv);
-eye(2,2)/Linv zeros(2,2)  zeros(2,3)  zeros(2,3);
zeros(3,2)  zeros(3,2)  -Ks*inv(Rload33)*Ksinv/Cload  eye(3,3)/Cload;
Trv_qd0/Ltrans  zeros(3,2)  -eye(3,3)/Ltrans  -Rtrans*eye(3,3)/Ltrans];
```

Form C matrix for 500% load:

```
Cov=[ [eye(10)];
   [zeros(3,4) Ks*inv(Rload33)*Ksinv zeros(3,3)]];
```

Step 4f. Define the A and C matrices for short circuit load.
Define the resistive load matrix for short circuit:

```
Rload33=[ 0.01    0    0;
          0    0.01    0;
          0     0    0.01];
```

Form A matrix for short circuit:

```
Asc=[ zeros(2,2)  cyc(2,2)/(3*Cinv)  zeros(2,3)  -Tri_qd0/(3*Cinv);
-eye(2,2)/Linv  zeros(2,2)  zeros(2,3)  zeros(2,3);
zeros(3,2)  zeros(3,2)  -Ks*inv(Rload33)*Ksinv/Cload  eye(3,3)/Cload;
Trv_qd0/Ltrans  zeros(3,2)  eye(3,3)/Ltrans  -Rtrans*eye(3,3)/Ltrans];
```

Form C matrix for short circuit:

```
Csc=[ [eye(10)];
[zeros(3,4) Ks*inv(Rload33)*Ksinv zeros(3,3)]];
```

2.6.4 The Voltage Controller and Current Limit Function

Implementation of the servovoltage controller in Simulink is straightforward. The servo-states equation is implemented in a *discrete state-space block*. The servo-compensator and stabilizing compensator gains are implemented using a *matrix gain* block. The block diagram of the voltage controller in Simulink is shown in Fig. 2.13.

The current limit function is implemented in the block called *CurrentLimit*. Its implementation is shown in Fig. 2.14. Detailed descriptions of each block can be found in the Appendix A.

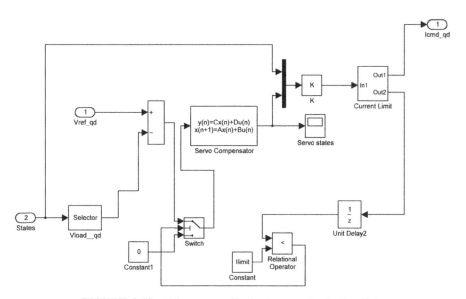

FIGURE 2.13 Voltage controller implementation in Simulink.

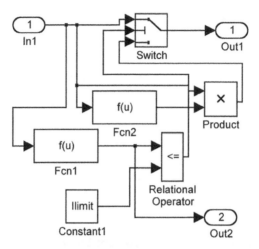

FIGURE 2.14   Current limit implementation in Simulink.

2.6.5   The Current Controller

Simulink block diagrams shown in Fig. 2.15 implement the discrete time sliding mode current controller. The equivalent control is computed based on Eq. 2.26.

The PWM command limit block implements the saturation function of the sliding mode current controller. Its implementation is shown in Fig. 2.16. It should be noted here that the PWM command limit function is similar to the current limit function explained earlier.

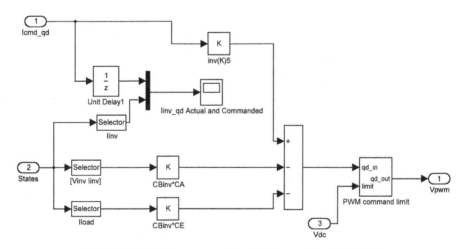

FIGURE 2.15   Current controllers implementation in Simulink.

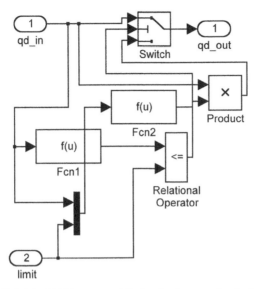

FIGURE 2.16   PWM command limiter implementation in Simulink.

2.7 STEP-BY-STEP INSTRUCTIONS ON USING THE SIMULINK SIMULATION

There are three files needed to run the simulation of the control algorithm. These files are as follows:

1. *ssmode.m*: This is a MATLAB script file that performs the following tasks:
   - Initializes constants and circuit parameters for the gains calculations and the Simulink model.
   - Calculates the servocontrol gains and the sliding mode control gains.
   - Computes the states equations of the plant for different types of loads.
     The full listing of this file is given in Appendix B. Detailed step-by-step explanations of the tasks performed in this file have been outlined in the previous sections.
2. *dsimservo.mdl*: This is a Simulink model file that simulates the control algorithm and the power converter plant. The description of each block diagram contained in this file is given in Appendix A. Detailed explanation of the controllers and the plant have been provided in the previous sections.
3. *sfunfft.m*: This is a Simulink S-Function block used to calculate the THD number of the output voltages. It uses a standard fast fourier transform algorithm to calculate the THD of a signal within a specified period. The full listing of

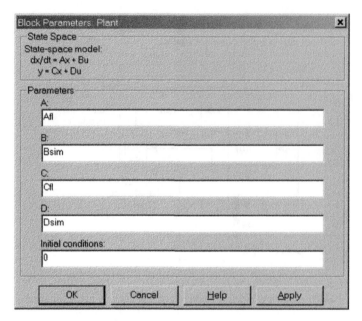

FIGURE 2.17   Dialog box for changing the load.

this file is given in Chapter 10, Section 10.2. Follow the steps below to run the simulation:

a. Open MATLAB program.

b. Change to the directory where the two files mentioned above reside using the command : *cd dir_name* on the MATLAB command prompt.

c. From the MATLAB command prompt, type *ssmode*. This will run the MAT-LAB script *ssmode.m,* which performs the necessary tasks explained above.

d. Open the *dsimservo.mdl* Simulink model by typing : *dsimservo* on the command prompt.

e. Run the simulation by selecting from the menu of the *dsimservo* Simulink window: *Simulation* → *Run.*

f. Open the scopes available on the *dsimservo.mdl* to see the signal waveforms. To change the load connected to the power converter, perform the following:

g. Double click the *Plant* block in the *dsimservo* Simulink window. This will open a dialogue box as shown in Fig. 2.17 that lets the user to change the A, B, C, and D parameters of the *Plant* state-space block.

h. Enter the A and C matrix corresponding to the desired load from Table 2.1 and hit the Apply button.

Note that the load can be changed during the simulation to simulate the transient effect.

TABLE 2.1   A and C Parameters for Different Loads

Types of Loads	A Matrix	C Matrix
100% load	A_{fl}	C_{fl}
No load	A_{nl}	C_{nl}
Single-phase 100% load	A_{p1}	C_{p1}
Two-phase 100% load	A_{p2}	C_{p2}
500% load	A_{ov}	C_{ov}
Short-circuit Load	A_{sc}	C_{sc}

Summarizes the A and C parameters for the different loads available.

To observe the effect of different optimization weighting on the performance of the control, do the following:

1. On the command prompt type: *edit ssmode*. This will open the script file in a MATLAB editor window.
2. Go to the line that specifies the weightings and make the changes

```
epsilon=1e-5;
state_W=0.2; % 80 kVA
```

3. Rerun the simulation.

As discussed in Section 2.6, the Simulink simulation diagram allows the generation of fundamental and harmonic frequencies for the voltages references. To change the amplitude and frequency of the harmonic, do the following:

Double click on one of the sine wave blocks. This will open a dialogue box as shown in Fig. 2.18.

Make the necessary changes to the amplitude and/or frequency parameters and hit the Apply button.

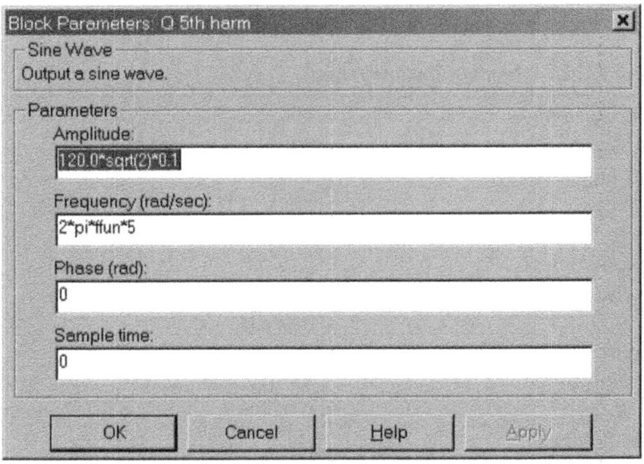

FIGURE 2.18   PWM command limiter implementation in Simulink.

2.8 SIMULATION AND EXPERIMENTAL RESULTS

This section presents a comparison between the simulation results obtained using the Simulink model *dsimservo* explained in the previous sections and the experimental results obtained in the actual hardware implementation. The experimental results were obtained from an 80-kVA UPS unit with system parameters shown in Table 2.2, which was also used to initialize the simulations. The DC bus voltage in the experimental work was obtained from a six-pulse thyristor-controlled rectifier in parallel with a 480-V battery system. The DSP control system used is based on the TMS320F240 fixed-point DSP with control timing diagram as given in Fig. 2.6. The PWM timing is calculated through a standard space vector PWM with switching frequency of 3.2 kHz ($T_{pwm} = T_s = 320\mu$ sec).

The simulation and experimental results presented in this chapter have been obtained using the proposed control strategy with the fifth and seventh harmonics being eliminated. The following cases will be investigated:

- Regulation under linear load
- Transient performance.
- Regulation under nonlinear load
- Short circuits at the output terminals

For cases where both simulation and experimental results exist, the two results will be compared.

2.8.1 Steady State Performance

Figure 2.19–2.24 show the steady-state waveforms of load currents, load voltages, and inverter voltages for different types of load both from simulations and from experimental results. The RMS values of the line-to-line load voltages for all cases are given in Table 2.3.

TABLE 2.2   System Parameters

DC bus voltages	
V_{DC}	540 V (nom.) 390 V (min)
AC output voltage	
V_{load}	208V (LL-RMS), 120 V(LN)
f	60 Hz
Inverter filters	
C_{inv}	540 μF
L_{inv}	300 μH
Delta–wye transformer	
L_{trans}	48 μH ($\approx$ 0.03 p.u)
R_{trans}	0.02 ohm
Output filter	
C_{grass}	90 μF

TABLE 2.3   Output Voltages Regulation

Load	Simulation			Experiments		
	Ph. A	Ph. B	Ph. C	Ph. A	Ph. B	Ph. C
No load	120.0	120.0	120.0	120.23	120.07	120.16
Full load	120.0	120.0	120.0	120.17	120.03	120.15
One-phase	119.1	121.3	119.7	120.59	119.59	120.21
Two-phase	118.7	120.3	121.0	120.67	119.97	119.72

It can be seen from Table 2.3 that there are slight differences in the numbers obtained from simulations and experimental results. This was caused mostly by the nonideality and parameter variations in the case of the experimental results. However, all the numbers show that the regulations of the load voltages are within the 2% specified by the project.

2.8.2   Transient Performance

Figures 2.25 and 2.26 show the simulation and experimental results, respectively, for 0–100% load transient. It can be seen that the waveforms of the voltages and currents for both cases look very similar. The voltage transients in both cases were characterized by the existence of dips in the voltages and followed by slight overshoots

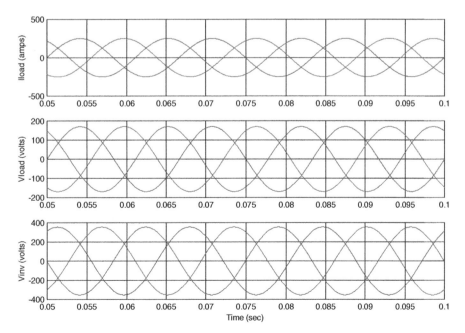

FIGURE 2.19   Simulation results for balanced full resistive load. **Top:** Load currents. **Middle:** Load voltages. **Bottom:** Inverter voltages.

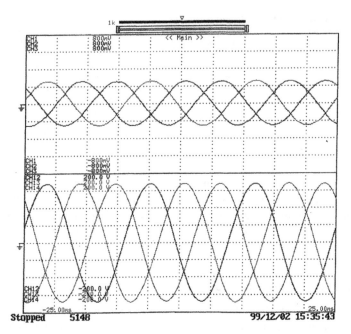

FIGURE 2.20 Experimental results for balanced full resistive load. **Top:** Load currents. **Bottom:** Load voltages.

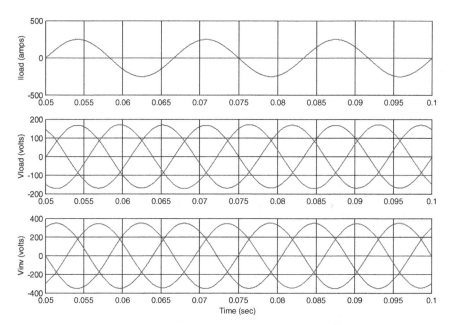

FIGURE 2.21 Simulation results for single-phase resistive load. **Top:** Load currents. **Middle:** Load voltages. **Bottom:** Inverter voltages.

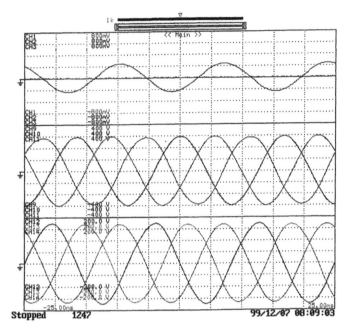

FIGURE 2.22 Experimental results for single-phase resistive load. **Top:** Load currents. **Middle:** Load voltages. **Bottom:** Inverter voltages.

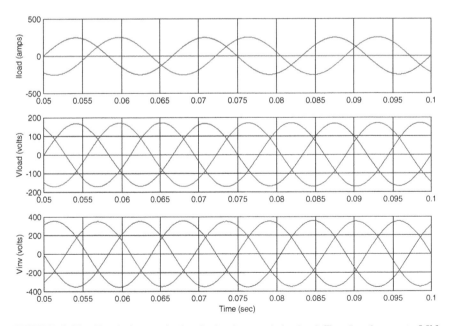

FIGURE 2.23 Simulation results for single-phase resistive load. **Top:** Load currents. **Middle:** Load voltages. **Bottom:** Inverter voltages.

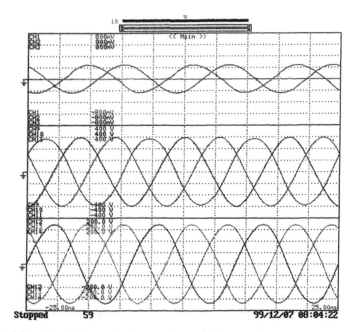

FIGURE 2.24 Simulation results for two-phase resistive load. **Top:** Load currents. **Middle:** Load voltages. **Bottom:** Inverter voltages.

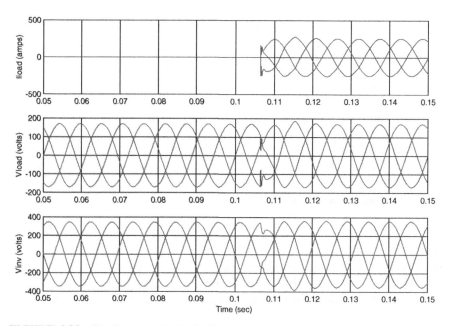

FIGURE 2.25 Simulation results for 0–100% load transient. **Top:** Load currents. **Middle:** Load voltages. **Bottom:** Inverter voltages.

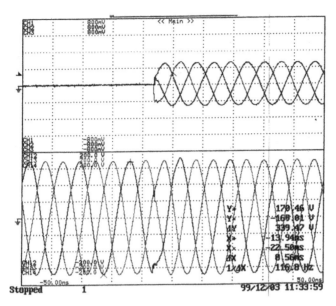

FIGURE 2.26 Experimental results for 0–100% load transient. **Top:** Load currents. **Bottom:** Load voltages.

exhibiting some fifth and seventh harmonic contents. The harmonics showed up as responses of the servoharmonic compensator to load transients. Note that the load transient lasts for only one cycle of the fundamental, which shows the fast response of the controller. Also, in any case the RMS load voltages variation were less than the 5% meeting the specification required by the project.

Figures 2.27 and 2.28 show the simulation and experimental results for the case when the load is transient from 100% to 0%. The experimental results show that the 100–0% load transient performance of the controller was satisfactory since it can be shown that the RMS voltage variation of the load voltages were less than the 5% specification.

2.8.3   Regulation Under Nonlinear Load

To verify the effectiveness of the controller in rejecting the disturbance caused by harmonics disturbances, the system needs to be tested under nonlinear load. However, simulation of nonlinear load is a time-consuming process, since it requires simulations of discontinuous systems. For this reason, the Simulink simulation provided in this chapter does not support this type of loads. Therefore, only experimental results are presented as shown in Fig. 2.29. An 80-kVA 2.75:1 nonlinear load was applied and the measured output voltage THD was 2.8% and the fifth and seventh were eliminated, thus proving the effectiveness of the RSP voltage controller.

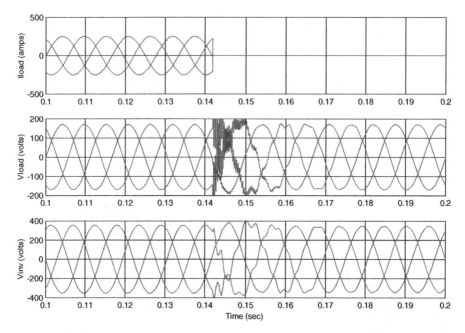

FIGURE 2.27 Nonrealistic simulation results for 100–0% load transient. **Top:** Load currents. **Middle:** Load voltages. **Bottom:** Inverter voltages.

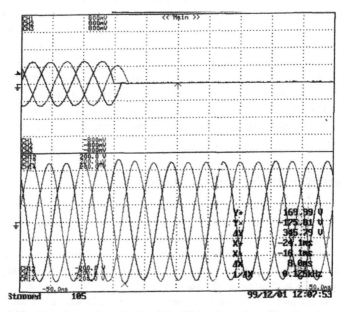

FIGURE 2.28 Experimental results for 100–0% load transient. **Top:** Load currents. **Bottom:** Load voltages.

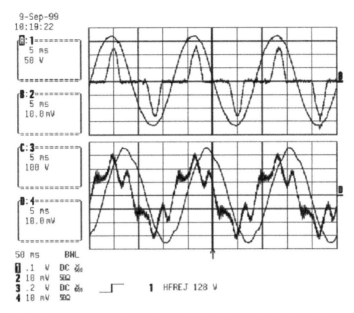

FIGURE 2.29   Steady-state nonlinear load: 100% 3:1 crest load. **Top:** Phase A of load voltages and load currents. **Bottom:** Phase A of inverter voltages and inverter currents.

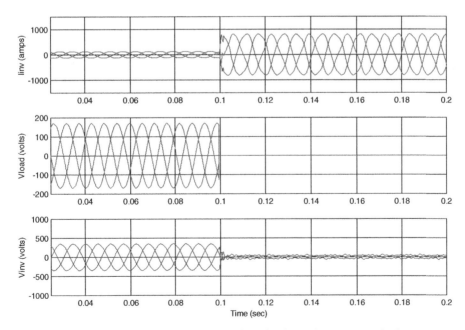

FIGURE 2.30   Simulation result: Short circuits at the output terminals.

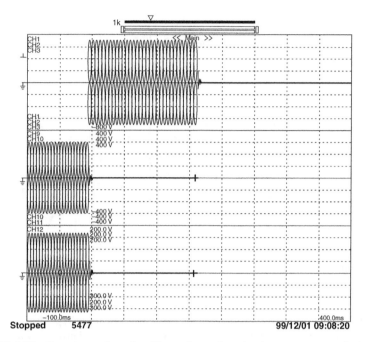

FIGURE 2.31   Experimental results: Three-phase shortcircuit on output terminals. **Top:** Inverter currents. **Middle:** Load voltages. **Bottom:** Inverter voltages.

2.8.4   Short Circuit at the Output Terminals

The final verification to the proposed control algorithm is for its ability to limit the inverter currents under short circuits at the output terminal. Figures 2.29 and 2.30 show the simulation and experimental results for this condition. In both cases, the current limit was set at the 300% level. In the experimental case, the inverter was shut down deliberately after 10 cycles of short-circuit condition. From these figures it can be seen that the discrete sliding mode controller provides a fast and minimal overshoot on the inverter currents.

CHAPTER 3

PARALLEL OPERATION OF INVERTERS IN DISTRIBUTED GENERATION SYSTEMS

3.1 INTRODUCTION

Recently, interest in distributed generation systems (DGS) is rapidly increasing, particularly for on-site generation. This interest is because larger power plants are economically unfeasible in many regions due to increasing system and fuel costs and more strict environmental regulations. In addition, recent technological advances in small generators, power electronics, and energy storage devices have provided a new opportunity for distributed energy resources at the distribution level; in particular, the incentive laws to utilize renewable energies have also encouraged a more decentralized approach to power delivery.

There exist various generation sources for DGS: conventional technologies (diesel or natural gas engines), emerging technologies (micro-turbines or fuel cells or energy storage devices), and renewable technologies (small wind turbines or solar/photovoltaics or small hydro-turbines). These DGS are used for applications to a standalone, a standby, a grid-interconnected, a cogeneration, peak shavings, and so on, and have many advantages such as environmental-friendly and modular electric generation, increased reliability, high power quality, uninterruptible service, cost savings, on-site generation, expandability, and so on.

Introduction of such DGS as combustion engine, small hydro, photovoltaic arrays, fuel cells, micro-turbines or battery energy storage systems, and cogenerations into power systems is very attractive for power utilities, customers, and societies because, for utility, DGS can help to improve power quality and power supply flexibility,

Integration of Green and Renewable Energy in Electric Power Systems. By A. Keyhani, M. N. Marwali, and M. Dai
Copyright © 2010 John Wiley & Sons, Inc.

maintain system stability, optimize the distribution system, provide the spinning reserve, and reduce the transmission and distribution cost and can be used to feed customers in the event of an outage in the line or in the primary substation or during scheduled interruptions. Also, for society, the renewable energy can significantly reduce the emission from the traditional power plants. So many utility companies are trying to construct small distribution stations combined with several DGS available at the regions, instead of large power plants.

Basically, these technologies are based on notably advanced power electronics because all DGS require power converters, PWM techniques, and electronic control units. That is, electrical power generated by all DGS is first converted into DC power, and then all the power fed to the DC distribution bus is again converted by DC to AC power converters into an AC power with fixed magnitude and frequency by control units using a digital signal processor (DSP). Therefore, in order to permit grid interconnection of asynchronous generation sources, advanced power electronic technologies are definitely required.

The DGS focuses on two applications: a standalone AC system and a grid-interconnection to the utility mains. In a standalone AC power supply, several DGS independently supply the loads with electrical power, like the parallel operation of the uninterruptible power supply (UPS) systems. In a grid-interconnected operation to AC mains, each DES is interconnected in parallel to the utility, and it directly provides power to AC mains in order to cover increased power required by the loads.

This chapter is concerned with the control strategy for the parallel operation of distributed generation systems (DGS) in a standalone AC power supply. In particular, the chapter focuses on proper power sharing of each DGS such as the real power, reactive power, and harmonic power in a standalone AC power supply, like the parallel operation of multiple UPS systems. First of all, good load-sharing should be maintained under both locally measurable voltages/currents and the wire impedance mismatches, voltage/current measurement error mismatches that significantly degrade the performance of load-sharing. Key features of the proposed control method are that it only uses locally measurable feedback signals (voltages/currents) and uses relatively low bandwidth data communication signals (respective real power and reactive power) between each generation system.

To ensure good load-sharing, the scheme combines two control methods: the droop control method and the average power control method. In this method, the sharing of real and reactive powers between each DGS is implemented by two independent control variables: power angle and inverter output voltage amplitude. Especially, the average power method is used in order to significantly reduce the sensitivity about voltage and current measurement error mismatches. This scheme guarantees good load-sharing of the fundamental components of the load currents.

The fundamental component load-sharing scheme above is designed around two feedback control loops: The inner loop is used for current control, while the outer one is for voltage control. A discrete-time sliding mode controller (DSMC) is applied as the current controller, and a robust servomechanism controller (RSC) is chosen to be the voltage controller. The DSMC is used in the current loop to limit the inverter current under overload condition because of the fast and no-overshoot response it provides. The RSC is adopted for voltage control due to its capability to perform zero steady-state tracking error under unknown load and eliminate harmonics of any specified frequencies with guaranteed system stability. Additionally, the RSC voltage control loop allows the use of a harmonic control droop scheme that ensures proper sharing of the harmonic components of the load currents.

In this study, two three-phase PWM inverters (600 kVA) operating in parallel are implemented by digital computer simulation using MATLAB/Simulink, and space vector pulse width modulation (SVPWM) is used as a PWM technique. Some simulation results under various operating conditions are also given.

3.2 DISTRIBUTED ENERGY SYSTEM DESCRIPTION

The configurations for two applications of DGS are shown in Fig. 3.1: a grid-interconnection and a standalone AC power supply. Figure 3.1a shows a schematic diagram of DGS used for a grid-interconnection application. As shown in Fig. 3.1a, DGS are interconnected to the conventional distribution lines in order to cover increased power required by the loads, so the distributed generation systems may spread around the distributed system that is connected to a grid system. In this application, DGS are connected to a medium–voltage network or a low–voltage network according to power ratings or voltage ratings available for the systems. Figure 3.1b shows the network structure of distributed generation systems used for a standalone AC power supply, and each distributed energy system supply power to the loads, like the parallel operation of UPS units in the emergency mode operation. As shown in Fig. 3.1b, this architecture may require each DGS to operate independently because the distance between DGS units may make data communication impractical, and the control should be based on those variables that can only be measured locally at the inverter. However, it was recently reported that data communication between units can easily be realized by the rapid advances in the field of communication.

Figure 3.2 shows a schematic diagram for data communication between a remote terminal unit (RTU) and an RTU for each DES to exchange power information (real power and reactive power) in a standalone AC power system.

A DGS circuit model in a standalone AC power system is shown in Fig. 3.3. This model consists of two three-phase PWM inverters running in parallel supporting two loads. The power converter of each DGS studied in this chapter is a DC/AC voltage source inverter, and each inverter is equipped with an L–C output filter and a Δ/Y transformer (600 kVA, Np:Ns = 245:208). In this figure, in order to simplify circuit analysis, generation sources such as combustion engine, small hydro, photovoltaic arrays, fuel cells, micro-turbines or battery energy storage systems, and so on, can be modeled as DC voltage sources (VDC1 and VDC2).

3.3 DGS CONTROL REQUIREMENTS

The droop technique has been widely used as a load-sharing scheme in conventional power system with multiple generators. In this droop method, the generators share

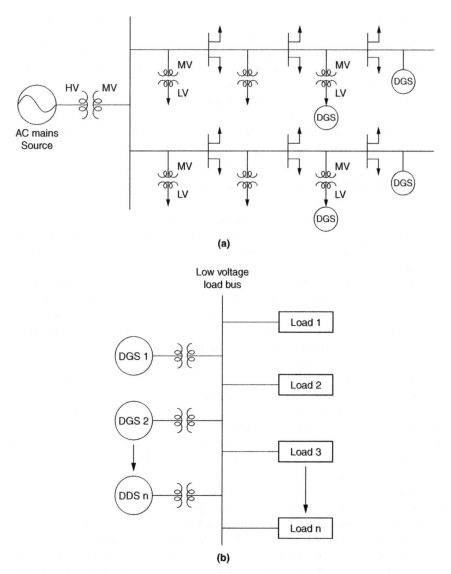

FIGURE 3.1 Configurations for two applications of DES. (**a**) Grid interconnection. (**b**) Standalone AC power supply.

the system load by drooping the frequency of each generator with the real power (P) delivered by the generator. This allows each generator to share changes in total load in a manner determined by its frequency droop characteristic and essentially utilizes the system frequency as a communication link between the generator control systems. Similarly, a droop in the voltage amplitude (V_{max}) with reactive power (Q) is used to ensure reactive power sharing.

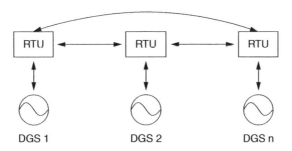

FIGURE 3.2   Control unit communications between DGS units.

This load-sharing technique is based on the power flow theory in an AC system, which states that the flow of the active power (P) and reactive power (Q) between two sources can be controlled by adjusting the power angle and the voltage magnitude of each system; that is, the active power flow (P) is predominantly controlled by the power angle, while the reactive power (Q) is predominantly controlled by the voltages magnitude. This theory is explained in Fig. 3.4. Figure 3.4 indicates critical variables for load-sharing control of paralleled power converters. The figure shows two inverters represented by two voltage sources connected to a load through line impedance represented by pure inductances L_1 and L_2 for simplified analysis purpose.

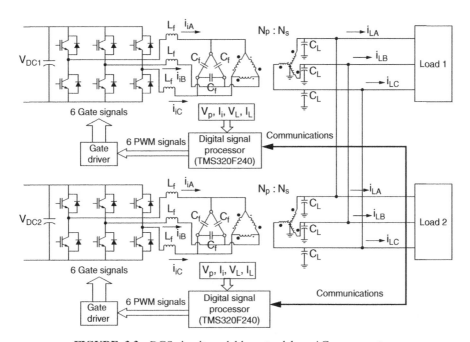

FIGURE 3.3   DGS circuit model in a standalone AC power system.

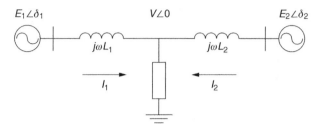

FIGURE 3.4   Two inverters connected to a load.

The complex power at the load due to inverter i is given by

$$S_i = P_i + jQ_i = V \cdot I_i^*,$$ (3.1)

where $i = 1, 2$ and I_i^* is the complex conjugate of the inverter i current and is given by

$$I_i^* = \left[\frac{E_i \cos \delta_i + jE_i \sin \delta_i - V}{j\omega L_i} \right]^*,$$ (3.2)

$$\therefore \quad S_i = V \left[\frac{E_i \cos \delta_i + jE_i \sin \delta_i - V}{j\omega L_i} \right]^*.$$ (3.3)

This gives the active and reactive power flowing from the ith inverter as

$$P_i = \frac{VE_i}{\omega L_i} \sin \delta_i,$$ (3.4)

$$Q_i = \frac{VE_i \cos \delta_i - V^2}{\omega L_i}.$$ (3.5)

From Eqs. (3.1)–(3.5), it can be seen that if δ_1 and δ_2 are small enough, then the real power flow is mostly influenced by the power angles δ_1 and δ_2, while the reactive power flow predominantly depends on the inverter output voltages E_1 and E_2. This means that to a certain extent the real and reactive power flow can be controlled independently. Since controlling the frequencies dynamically controls the power angles, the real power flow control can be equivalently achieved by controlling the frequencies of the voltages generated by the inverters.

Therefore, as mentioned above, the power angle and the inverter output voltage magnitude are critical variables that can directly control the real and reactive power flow for proper load-sharing of power converters connected in parallel.

Similarly, the above control theory can be applied to parallel operation of distributed energy systems in a standalone AC power supply application. In general, there is a large distance between inverter output and load bus, so each DGS is required to operate independently by using only locally measurable voltages/currents information. In addition, there is also a long distance between DGS units, so data communication between DGS units may be impractical. However, in recent years, data communication between DGS units has easily been implemented by the rapid advances in the field of communication.

Therefore, the chapter considers the parallel operation of power converters in a standalone AC system under signal communications between units in order to ensure exact load-sharing of each DGS unit. In particular, this research focuses on proper load-sharing between each unit, and the load-sharing should be also guaranteed under the wire impedance mismatches, voltage/current measurement error mismatches, and the interconnection tie-line impedance effect that can heavily affect the performance of load-sharing.

Control constraints considered in this chapter are summarized as follows:

- Locally measurable feedback signals (voltages/currents)
- Data communications between each DGS about real power and reactive power
- Wire impedance mismatches between inverter output and load bus
- Voltage/current sensor measurement error mismatches
- Tie-line impedance between loads

To overcome the above control constraints, a new droop technique using both average power method and harmonic droop control is proposed.

3.4 DISTRIBUTED GENERATION SYSTEM MODELING

Figure 3.5 illustrates a circuit model of each DGS studied in this chapter, and each DGS consists of a DC voltage source, a three-phase PWM inverter, an L–C output filter, a Δ/Y transformer, and a load. Of course, a DC voltage source may come from one of various generation sources: combustion engine, small hydro, photovoltaic arrays, fuel cells, micro-turbines, or battery energy storage systems.

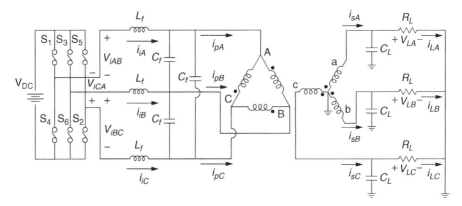

FIGURE 3.5 Each DGS circuit model.

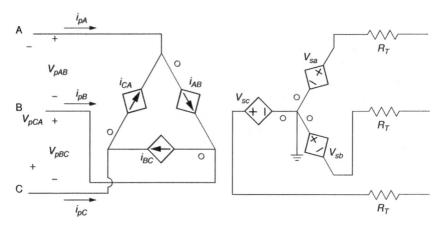

FIGURE 3.6 Δ/Y transformer model.

Particularly, the three-phase (Δ/Y type) transformer is further modeled using a controlled voltage source, a controlled current source, and equivalent phase impedances with the resistor R_T and the inductor L_T as shown in Fig. 3.6. The power rating of each transformer is 600 kVA, and turn ratio between the primary side (N_p) and secondary side (N_s) is 245:208. The primary side and secondary side voltages are 980 V (RMS, line to line) and 832 V (RMS, line to line), respectively.

The circuit system defined in Figs. 3.5 and 3.6 uses the following quantities to describe its behavior. The inverter output line-to-line voltage is represented by the vector $\mathbf{V}_i = [V_{iAB} \ V_{iBC} \ V_{iCA}]^T$. The three-phase inverter output currents are i_{iA}, i_{iB}, and i_{iC}. Based on these currents, a vector can be defined as $\mathbf{I}_i = [i_{iAB} \ i_{iBC} \ i_{iCA}]^T = [i_{iA} - i_{iB} \ i_{iB} - i_{iC} \ i_{iC} - i_{iA}]^T$. The transformer primary side line-to-line voltage is represented by the vector $\mathbf{V}_p = [V_{pAB} \ V_{pBC} \ V_{pCA}]^T$. The transformer primary side is Δ-connected and the line current vector is defined as $\mathbf{I}_p = [I_{pA} \ I_{pB} \ I_{pC}]^T$. The transformer secondary side is Y-connected and the phase voltage and current vectors are represented by $\mathbf{V}_s = [V_{sa} \ V_{sb} \ V_{sc}]^T$ and $\mathbf{I}_s = [I_{sa} \ I_{sb} \ I_{sc}]^T$, respectively. At the output terminal, the load voltage and current vectors can be represented by $\mathbf{V}_L = [V_{La} \ V_{Lb} \ V_{Lc}]^T$ and $\mathbf{I}_L = [I_{La} \ I_{Lb} \ I_{Lc}]^T$, respectively.

The three-phase (Δ/Y type) transformer has N_p turns in the primary side windings and N_s turns in the secondary side windings. Based on Fig. 3.6, the voltage relation between the two sides of the transformer can be described by

$$V_{sa} = \frac{N_s}{N_p} V_{pAB}, \qquad V_{sb} = \frac{N_s}{N_p} V_{pBC}, \qquad V_{sc} = \frac{N_s}{N_p} V_{pCA}. \qquad (3.6)$$

Similarly, the current relation is

$$i_{AB} = \frac{N_s}{N_p} i_{sa}, \qquad i_{BC} = \frac{N_s}{N_p} i_{sb}, \qquad i_{CA} = \frac{N_s}{N_p} i_{sc}. \qquad (3.7)$$

From Fig. 3.6, it also can be observed that

$$i_{pA} = i_{AB} - i_{CA}, \qquad i_{pB} = i_{BC} - i_{AB}, \qquad i_{pC} = i_{CA} - i_{BC}. \qquad (3.8)$$

On the primary side of the transformer, the L–C filter yields the following current equations:

$$i_{iA} + C_f \frac{dV_{pCA}}{dt} = C_f \frac{dV_{pAB}}{dt} + i_{pA},$$

$$i_{iB} + C_f \frac{dV_{pAB}}{dt} = C_f \frac{dV_{pBC}}{dt} + i_{pB}, \qquad (3.9)$$

$$i_{iC} + C_f \frac{dV_{pBC}}{dt} = C_f \frac{dV_{pCA}}{dt} + i_{pC}.$$

From Eqs. (3.7)–(3.9), it is easy to derive

$$\frac{dV_{pAB}}{dt} = \frac{i_{iAB}}{3C_f} + \frac{1}{3C_f} \frac{N_s}{N_p} (2i_{sa} - i_{sb} - i_{sc}),$$

$$\frac{dV_{pBC}}{dt} = \frac{i_{iBC}}{3C_f} + \frac{1}{3C_f} \frac{N_s}{N_p} (-i_{sa} + 2i_{sb} - i_{sc}), \qquad (3.10)$$

$$\frac{dV_{pCA}}{dt} = \frac{i_{iCA}}{3C_f} + \frac{1}{3C_f} \frac{N_s}{N_p} (-i_{sa} - i_{sb} + 2i_{sc}).$$

Rewrite Eq. (3.10) into matrix form:

$$\frac{d\mathbf{V}_p}{dt} = \frac{1}{3C_f} \mathbf{I}_i - \frac{1}{3C_f} \mathbf{T}_i \mathbf{I}_s, \qquad (3.11)$$

where

$$\mathbf{T}_i = \frac{N_s}{N_p} \begin{bmatrix} 2 & -1 & -1 \\ -1 & 2 & -1 \\ -1 & -1 & 2 \end{bmatrix}. \qquad (3.12)$$

Equation (3.6) can be rewritten as

$$\mathbf{V}_s = \begin{bmatrix} V_{sa} \\ V_{sb} \\ V_{sc} \end{bmatrix} = \frac{N_s}{N_p} \begin{bmatrix} 1 & 0 & 0 \\ 0 & 1 & 0 \\ 0 & 0 & 1 \end{bmatrix} \begin{bmatrix} V_{pAB} \\ V_{pBC} \\ V_{pCA} \end{bmatrix} = \mathbf{T}_v \mathbf{V}_p. \qquad (3.13)$$

It is easy to write the L–C filter voltage equations as follows:

$$L_f \frac{di_{iA}}{dt} - L_f \frac{di_{iB}}{dt} = V_{iAB} - V_{pAB},$$

$$L_f \frac{di_{iB}}{dt} - L_f \frac{di_{iC}}{dt} = V_{iBC} - V_{pBC}, \tag{3.14}$$

$$L_f \frac{di_{iC}}{dt} - L_f \frac{di_{iA}}{dt} = V_{iCA} - V_{pCA}.$$

Rewrite Eq. (3.14) into matrix form:

$$\frac{d\mathbf{I}_i}{dt} = \frac{1}{L_f} \mathbf{V}_i - \frac{1}{L_f} \mathbf{V}_p. \tag{3.15}$$

The load current equation can be written as

$$\frac{d\mathbf{V}_L}{dt} = \frac{1}{C_L} \mathbf{I}_s - \frac{1}{C_L} \mathbf{I}_L. \tag{3.16}$$

The voltage equation of the secondary side circuit is

$$\frac{d\mathbf{I}_s}{dt} = -\frac{R_T}{L_T} \mathbf{I}_s - \frac{1}{L_T} \mathbf{T}_v \mathbf{V}_p - \frac{1}{L_T} \mathbf{V}_L. \tag{3.17}$$

Equations (3.11), (3.15), (3.16), and (3.17) are the four state equations for each DGS circuit model—or, say, the control plant of the proposed feedback system in this chapter. The state variables of the system are $\mathbf{V}_p$, $\mathbf{I}_i$, $\mathbf{V}_L$, and $\mathbf{I}_s$, the control input is the inverter output line-to-line voltage $\mathbf{V}_i$, and the disturbance is the load current $\mathbf{I}_L$.

The three-phase system represented by the above state space model can be transformed from the *abc* reference frame into stationary *qd*0 reference frame using Eq. (3.18).

$$\mathbf{f}_{qd0} = \mathbf{K}_s \mathbf{f}_{abc}, \tag{3.18}$$

where

$$K_s = \frac{2}{3} \begin{bmatrix} 1 & -1/2 & -1/2 \\ 0 & -\sqrt{3}/2 & \sqrt{3}/2 \\ 1/2 & 1/2 & 1/2 \end{bmatrix}, \quad \mathbf{f}_{qd0} = [f_q \ f_d \ f_0]^T, \quad \mathbf{f}_{abc} = [f_a \ f_b \ f_c]^T,$$

and f denotes either a voltage or a current variable.

Rewrite the previous state equations [(3.11), (3.15)–(3.17)] into the stationary $qd0$ reference frame defined above:

$$\frac{d\mathbf{V}_{pqd}}{dt} = \frac{1}{3C_f}\mathbf{I}_{iqd} - \frac{1}{3C_f}\mathbf{T}_{iqd0}\mathbf{I}_{sqd0}, \tag{3.19}$$

$$\frac{d\mathbf{I}_{iqd}}{dt} = \frac{1}{L_f}\mathbf{V}_{iqd} - \frac{1}{L_f}\mathbf{V}_{pqd}, \tag{3.20}$$

$$\frac{d\mathbf{V}_{Lqd0}}{dt} = \frac{1}{C_L}\mathbf{I}_{sqd0} - \frac{1}{C_L}\mathbf{I}_{Lqd0}, \tag{3.21}$$

$$\frac{d\mathbf{I}_{sqd0}}{dt} = -\frac{R_T}{L_T}\mathbf{I}_{sqd0} - \frac{1}{L_T}\mathbf{T}_{vqd0}\mathbf{V}_{pqd} - \frac{1}{L_T}\mathbf{V}_{Lqd0}, \tag{3.22}$$

where

$$\mathbf{T}_{iqd0} = [\mathbf{K}_s\mathbf{T}_i\mathbf{K}_s^{-1}]_{row\ 1,\ 2} = \frac{N_s}{N_p}\frac{3}{2}\begin{bmatrix} 2 & 0 & 0 \\ 0 & 2 & 0 \end{bmatrix}$$

and

$$\mathbf{T}_{vqd0} = [\mathbf{K}_s\mathbf{T}_v\mathbf{K}_s^{-1}]_{col\ 1,\ 2} = \frac{N_s}{N_p}\begin{bmatrix} 1 & 0 \\ 0 & 1 \\ 0 & 0 \end{bmatrix}.$$

All the above vectors with qd subscriptions are two dimensional vectors where the 0-axis elements are not included. Because the transformer is Δ/Y type with the neutral point of the secondary side grounded, the secondary side circuit may carry unbalanced three-phase current (e.g., load unbalanced fault can cause 0-axis current). The transient behavior of the 0-axis circuit is uncontrollable in that the circuit is not accessible for the control inputs V_{iq} and V_{id}. However, due to the existence of R_T, the circuit is asymptotically stable. Therefore, the overall system is stable according to the control theory that a linear time invariant system is stable if its uncontrollable modes are stable.

3.5 CONTROL SYSTEM DESIGN

3.5.1 Overall System Structure

The entire closed-loop system structure can be illustrated by the following block diagram:

In Fig. 3.7, RSC is the robust servomechanism controller, DSMC is the discrete-time sliding mode controller, SVPWM is a three-phase space vector pulse width modulation inverter, $\mathbf{I}^*_{cmd,qd}$ is the current command signal, $\mathbf{I}_{cmd,qd}$ is the limited

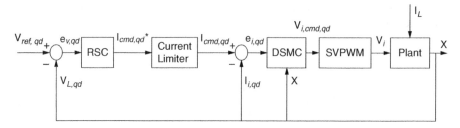

FIGURE 3.7   Overall control system structure.

current command, $V_{i,cmd,qd}$ is the inverter voltage command, and V_i is the true inverter output voltage. The transformations between quantities under qd reference frame (with qd subscription) and those under abc reference frame are not explicitly shown in the diagram.

As shown in Fig. 3.7, the overall system has two feedback loops. The inner loop is the current control loop where the regulator is the DSMC and the outer loop is the voltage control loop where the regulator is the RSC. The RSC and DSMC design will be described as follows.

3.5.2   Discrete-Time Sliding Mode Controller

Figure 3.8 shows the block diagram of the discrete-time sliding mode controller (DSMC). As shown in Fig. 3.8, the error $e_{i,qd}$ is used for the input signal of current controller, and the controller generates inverter output voltage command $V_{i,cmd,qd}$ as a control signal.

For the control of the inverter current in the system, only the subsystem represented by (3.19) and (3.20) needs to be considered, where I_{sqd} acts as a disturbance. Rewrite the equations in state-space form:

$$\dot{X}_1 = A_1 X_1 + B_1 u + E_1 d_1,$$
$$y_1 = C_1 X_1,$$

(3.23)

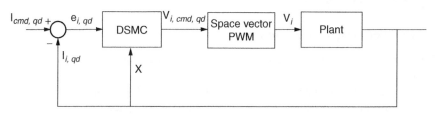

FIGURE 3.8   Discrete-time sliding mode controller.

where

$$\mathbf{X}_1 = \begin{bmatrix} \mathbf{V}_{pqd} \\ \mathbf{I}_{iqd} \end{bmatrix}, \quad \mathbf{A}_1 = \begin{bmatrix} 0_{2\times2} & \dfrac{1}{3C_f} I_{2\times2} \\ -\dfrac{1}{L_f} I_{2\times2} & 0_{2\times2} \end{bmatrix}, \quad \mathbf{B}_1 = \begin{bmatrix} 0_{2\times2} \\ \dfrac{1}{L_f} I_{2\times2} \end{bmatrix},$$

$$\mathbf{E}_1 = \begin{bmatrix} -\dfrac{1}{3C_f} \mathbf{T}_{iqd} \\ 0_{2\times2} \end{bmatrix}_{4\times2}, \quad \mathbf{C}_1 = \begin{bmatrix} 0 & 0 & 1 & 0 \\ 0 & 0 & 0 & 1 \end{bmatrix}, \quad \mathbf{u} = \begin{bmatrix} V_{iq} \\ V_{id} \end{bmatrix}, \quad \mathbf{d}_1 = \begin{bmatrix} i_{sq} \\ i_{sd} \end{bmatrix},$$

and

$$\mathbf{T}_{iqd} = \mathbf{T}_{iqd0,\, col\ 1,2} = \frac{N_s}{N_p} \frac{3}{2} \begin{bmatrix} 1 & \sqrt{3} \\ -\sqrt{3} & 1 \end{bmatrix}.$$

The 0-axis quantities are not included in the equation because they have no impact on the primary side of the system. The discrete form of the system in Eq. (3.23) can be represented by

$$\mathbf{X}_1(k+1) = \mathbf{A}_1^* \mathbf{X}_1(k) + \mathbf{B}_1^* \mathbf{u}(k) + \mathbf{E}_1^* \mathbf{d}_1(k),$$
$$\mathbf{y}_1(k) = \mathbf{C}_1 \mathbf{X}_1(k), \tag{3.24}$$
$$\mathbf{e}(k) = \mathbf{y}_{ref}(k) - \mathbf{y}_1(k),$$

where $\mathbf{A}_1^* = e^{A_1 T_s}$, $\mathbf{B}_1^* = \int_0^{T_s} e^{A_1(T_s - \tau)} \mathbf{B}_1 d\tau$, and $\mathbf{E}_1^* = \int_0^{T_s} e^{A_1(T_s - \tau)} \mathbf{E}_1 d\tau$, assuming sampling period T_s.

It is desired to control the output $\mathbf{y}_1(k)$ to follow the reference $\mathbf{y}_{ref}(k)$. For this purpose we can choose a sliding mode manifold in the form of

$$\mathbf{s}(k) = \mathbf{C}_1 \mathbf{X}_1(k) - \mathbf{y}_{ref}(k), \tag{3.25}$$

that is, the tracking error, such that when the discrete-time sliding mode exists, the output $\mathbf{y}_1(k)$ tends to the reference $\mathbf{y}_{ref}(k)$. The discrete-time sliding mode can be reached if the control input $u(k)$ is designed to be the solution of

$$\mathbf{s}(k+1) = \mathbf{C}_1 \mathbf{A}_1^* \mathbf{X}_1(k) + \mathbf{C}_1 \mathbf{B}_1^* \mathbf{u}(k) + \mathbf{E}_1^* \mathbf{d}_1(k) - \mathbf{y}_{ref}(k+1) = 0. \tag{3.26}$$

The control law satisfying (23) is called 'equivalent control' and is given by

$$\mathbf{u}_{eq}(k) = \left(\mathbf{C}_1 \mathbf{B}_1^*\right)^{-1} \left(\mathbf{I}_{cmd,qd}(k) - \mathbf{C}_1 \mathbf{A}_1^* \mathbf{X}_1(k) - \mathbf{C}_1 \mathbf{E}_1^* \mathbf{d}_1(k)\right) \tag{3.27}$$

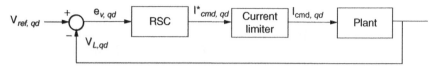

FIGURE 3.9   Voltage controller using RSC.

If the control is limited by $\|\mathbf{u}(k)\| \leq u_0$ then the following modified control law can be applied

$$\mathbf{u}(k) = \begin{cases} \mathbf{u}_{eq}(k) & \text{for } \|\mathbf{u}_{eq}(k)\| \leq u_0, \\[4mm] \dfrac{u_0}{\|\mathbf{u}_{eq}(k)\|}\mathbf{u}_{eq}(k) & \text{for } \|\mathbf{u}_{eq}(k)\| > u_0. \end{cases} \tag{3.28}$$

The control voltage limit $\mathbf{u}_0$ is determined by the SVPWM inverter. With control law (3.28), discrete-time sliding mode can be reached after a finite number of steps.

3.5.3   Voltage Controller using Discrete RSC

In Fig. 3.9, a discrete-time RSC is used for outer loop regulation. Once the solution to the given plant, reference signal, and disturbance exists, the controller can be designed analytically. The voltage control loop designed in this chapter is based on the discrete form of the technique developed in Davison's work. To design the load voltages controller, let's first consider the entire plant system.

Since the dynamics of the DSMC is included in the inner loop, its model has to be included together with the original plant to form the control plant for the RSC.

The original plant is

$$\dot{\mathbf{X}} = \mathbf{AX} + \mathbf{Bu} + \mathbf{Ed}, \tag{3.29}$$

where

$$\mathbf{X} = \begin{bmatrix} \mathbf{V}_{pqd} \\ \mathbf{I}_{iqd} \\ \mathbf{V}_{Lqd} \\ \mathbf{I}_{sqd} \end{bmatrix}, \quad \mathbf{A} = \begin{bmatrix} 0_{2\times2} & \dfrac{1}{3C_f}I_{2\times2} & 0_{2\times2} & -\dfrac{1}{3C_f}\mathbf{T}_{iqd} \\[3mm] -\dfrac{1}{L_f}I_{2\times2} & 0_{2\times2} & 0_{2\times2} & 0_{2\times2} \\[3mm] 0_{2\times2} & 0_{2\times2} & 0_{2\times2} & \dfrac{1}{C_L}I_{2\times2} \\[3mm] \dfrac{1}{L_T}\mathbf{T}_{vqd} & 0_{2\times2} & -\dfrac{1}{L_T}I_{2\times2} & -\dfrac{R_T}{L_T}I_{2\times2} \end{bmatrix}_{8\times8},$$

$$\mathbf{B} = \begin{bmatrix} 0_{2\times2} \\ \dfrac{1}{L_f} I_{2\times2} \\ 0_{2\times2} \\ 0_{2\times2} \end{bmatrix}_{8\times2}, \quad \mathbf{E} = \begin{bmatrix} 0_{2\times2} \\ 0_{2\times2} \\ -\dfrac{1}{C_L} I_{2\times2} \\ 0_{2\times2} \end{bmatrix}_{8\times2}, \quad \mathbf{u} = \begin{bmatrix} V_{iq} \\ V_{id} \end{bmatrix}, \quad \mathbf{d} = \begin{bmatrix} i_{Lq} \\ i_{Ld} \end{bmatrix},$$

and

$$\mathbf{T}_{vqd} = \mathbf{T}_{vqd0,\ row\ 1,\ 2} = \frac{N_s}{N_p} \frac{1}{2} \begin{bmatrix} 1 & -\sqrt{3} \\ \sqrt{3} & 1 \end{bmatrix}.$$

Since the 0-axis quantities of the secondary side are uncontrollable, they are not included. Given the sampling period T_s, the discrete form of the plant is

$$\mathbf{X}(k+1) = \mathbf{A}^*\mathbf{X}(k) + \mathbf{B}^*\mathbf{u}(k) + \mathbf{E}^*\mathbf{d}(k), \tag{3.30}$$

where $\mathbf{A}^* = e^{\mathbf{A}T_s}$, $\mathbf{B}^* = \int_0^{T_s} e^{\mathbf{A}(T_s-\tau)}\mathbf{B}\,d\tau$, and $\mathbf{E}^* = \int_0^{T_s} e^{\mathbf{A}(T_s-\tau)}\mathbf{E}\,d\tau$.

After the dynamics of the DSMC is included, the overall plant for the RSC is

$$\begin{aligned} \mathbf{X}(k+1) &= \mathbf{A}_d\mathbf{X}(k) + \mathbf{B}_d\mathbf{u}_1(k) + \mathbf{E}^*\mathbf{d}(k), \\ \mathbf{y}(k) &= \mathbf{C}_d\mathbf{X}(k), \end{aligned} \tag{3.31}$$

where

$$\mathbf{A}_d = \mathbf{A}^* - \mathbf{B}^*\left(\mathbf{C}_1\mathbf{B}_1^*\right)^{-1}\mathbf{C}_1\left(\mathbf{A}_1^*\mathbf{C}_{11} - \mathbf{E}_1^*\mathbf{C}_{12}\right), \qquad \mathbf{B}_d = \mathbf{B}^*\left(\mathbf{C}_1\mathbf{B}_1^*\right)^{-1},$$

$$\mathbf{C}_{11} = \begin{bmatrix} I_{2\times2} & 0_{2\times2} & 0_{2\times2} & 0_{2\times2} \\ 0_{2\times2} & I_{2\times2} & 0_{2\times2} & 0_{2\times2} \end{bmatrix}, \qquad \mathbf{C}_{12} = \begin{bmatrix} 0_{2\times2} & 0_{2\times2} & 0_{2\times2} & I_{2\times2} \end{bmatrix},$$

$$\mathbf{C}_d = \begin{bmatrix} 0_{2\times2} & 0_{2\times2} & I_{2\times2} & 0_{2\times2} \end{bmatrix}, \qquad \mathbf{u}_1(k) = \mathbf{I}_{cmd,qd}(k),$$

Assuming the tracking/disturbance poles are $\pm j\omega_1, \pm j\omega_3, \pm j\omega_5, \ldots$ (i.e., representing sinusoidal signals with fundamental frequency ω_1 and harmonic frequencies $\omega_3, \omega_5, \ldots$), the RSC can be designed as follows. If the tracking/disturbance poles to be considered are $\pm j\omega_1, \pm j\omega_3, \pm j\omega_5$, and $\pm j\omega_7$, the servocompensator is

$$\dot{\boldsymbol{\eta}} = \mathbf{A}_c\boldsymbol{\eta} + \mathbf{B}_c\mathbf{e}_{vqd}, \tag{3.32}$$

where $\mathbf{e}_{vqd} = \mathbf{V}_{ref,qd} - \mathbf{V}_{Lqd}$ is the load voltage tracking error,

$$
\mathbf{A}_c = \begin{bmatrix}
0 & I_{2\times2} & & & & & & \\
-\omega_1 I_{2\times2} & 0 & & & & & & \\
& & 0 & I_{2\times2} & & & & \\
& & -\omega_3 I_{2\times2} & 0 & & & & \\
& & & & 0 & I_{2\times2} & & \\
& & & & -\omega_5 I_{2\times2} & 0 & & \\
& & & & & & 0 & I_{2\times2} \\
& & & & & & -\omega_7 I_{2\times2} & 0
\end{bmatrix}_{16\times16},
$$

and

$$
\mathbf{B}_c = \begin{bmatrix} 0_{2\times2} & I_{2\times2} & 0_{2\times2} & I_{2\times2} & 0_{2\times2} & I_{2\times2} & 0_{2\times2} & I_{2\times2} \end{bmatrix}^T.
$$

Note that $\mathbf{A}_c$ has the same poles as the given tracking/disturbance poles. Rewrite Eq. (3.32) in discrete form:

$$
\boldsymbol{\eta}(k+1) = \mathbf{A}_c^*\boldsymbol{\eta}(k) + \mathbf{B}_c^*\mathbf{e}_{vqd}(k), \tag{3.33}
$$

where $\mathbf{A}_c^* = e^{\mathbf{A}_c T_s}$ and $\mathbf{B}_c^* = \int_0^{T_s} e^{\mathbf{A}_c(T_s-\tau)}\mathbf{B}_c\,d\tau$.

An augmented system combining both the plant and the servocompensator can be written as

$$
\hat{\mathbf{X}}(k+1) = \hat{\mathbf{A}}\hat{\mathbf{X}}(k) + \hat{\mathbf{B}}\mathbf{u}_1(k) + \hat{\mathbf{E}}_1\mathbf{d}(k) + \hat{\mathbf{E}}_2\mathbf{y}_{ref}(k), \tag{3.34}
$$

where

$$
\hat{\mathbf{X}}(k) = \begin{bmatrix} \mathbf{X}(k) \\ \boldsymbol{\eta}(k) \end{bmatrix}, \qquad
\hat{\mathbf{A}} = \begin{bmatrix} \mathbf{A}_d & 0 \\ -\mathbf{B}_c^*\mathbf{C}_d & \mathbf{A}_c^* \end{bmatrix}, \qquad
\hat{\mathbf{B}} = \begin{bmatrix} \mathbf{B}_d \\ 0 \end{bmatrix},
$$

$$
\hat{\mathbf{E}}_1 = \begin{bmatrix} \mathbf{E}^* \\ 0 \end{bmatrix}, \qquad
\hat{\mathbf{E}}_2 = \begin{bmatrix} 0 \\ \mathbf{B}_c^* \end{bmatrix},
$$

$$
\mathbf{u}_1(k) = \mathbf{I}_{cmd,qd}(k), \qquad \mathbf{d}(k) = \mathbf{I}_{Lqd}(k), \qquad \mathbf{y}_{ref}(k) = \mathbf{V}_{ref,qd}(k).
$$

Here, $\mathbf{u}_1$ is the control variable, $\mathbf{y}_{ref}$ is the reference input, and $\mathbf{d}$ is the disturbance as defined in the previous section.

Figure 3.10 shows the block diagram of RSC with servocompensator and stabilizing compensator. The stabilizing compensator, which yields the control signal $\mathbf{u}$ in Eq. (3.34), ensures the stability of the overall system including the modes in the plant as well as the servo-compensator and desirable performance of the system through a

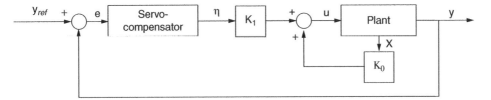

FIGURE 3.10   Feedback system with RSC.

feedback gain **K** which minimizes the linear quadratic performance index (i.e., the optimization criterion).

The final step is to design the control ($u_1(k)$) below:

$$\mathbf{u}_1(k) = \mathbf{K}\hat{\mathbf{X}}(k) = \begin{bmatrix} \mathbf{K}_0 & \mathbf{K}_1 \end{bmatrix} \begin{bmatrix} \mathbf{X}(k) \\ \boldsymbol{\eta}(k) \end{bmatrix} = \mathbf{K}_0\mathbf{X}(k) + \mathbf{K}_1\boldsymbol{\eta}(k). \tag{3.35}$$

The optimization criterion that minimizes the discrete linear quadratic performance index is as follows:

$$J_\varepsilon = \sum_{k=0}^{\infty} \hat{\mathbf{X}}^T(k)\mathbf{Q}\hat{\mathbf{X}}(k) + \varepsilon\mathbf{u}^T(k)\mathbf{u}(k), \tag{3.36}$$

where **Q** is a symmetrical positive-definite matrix and $\varepsilon > 0$ is a small number, both of which should be selected by the controller designer according to the application. The feedback gain **K** can be solved using MATLAB function *dlqr()*, which solves the algebraic Riccati equation for system (3.36).

3.5.4   Current Limit and Control Saturation Handling

The current command signal needs to be limited to perform overload protection. The algorithm of the current limiter is given as

$$\mathbf{I}_{cmd}(k) = \begin{cases} \mathbf{I}^*_{cmd}(k) & \text{for } \left\| \mathbf{I}^*_{cmd}(k) \right\| \leq I_{\max}, \\[2ex] \dfrac{I_{\max}}{\left\| \mathbf{I}^*_{cmd}(k) \right\|}\mathbf{I}^*_{cmd,qd}(k) & \text{for } \left\| \mathbf{I}^*_{cmd}(k) \right\| > I_{\max}. \end{cases} \tag{3.37}$$

To prevent servocompensator states, which are related to the current command, from growing while the current command is saturated, the following strategy can be applied. Rewrite the servocompensator equation as

$$\boldsymbol{\eta}(k+1) = \mathbf{A}^*_c\boldsymbol{\eta}(k) + \mathbf{B}^*_c\mathbf{e}_1(k), \tag{3.38}$$

where

$$\mathbf{e}_1(k) = \begin{cases} \mathbf{e}_v(k) & \text{if } \left\| \mathbf{I}^*_{cmd} \right\| \leq I_{\max}, \\ 0 & \text{otherwise.} \end{cases}$$

3.6 PROPOSED LOAD SHARING CONTROL ALGORITHM

3.6.1 Combined Droop Control Method and Average Power Control Method

In order to guarantee exact load-sharing of the real power (P) and the reactive power (Q) between DGS units, the conventional droop method should be modified by a new control algorithm. To do this, a combination of the droop method, the average power control method, and harmonic sharing control loop is proposed for load-sharing of paralleled inverters. Figure 3.11 shows a Simulink model of an average power generation block used for the proposed droop method. As shown in this figure, an average real power $(P_{avg\_qd} = (P_{1\_qd} + P_{2\_qd})/2)$ and reactive power $(Q_{avg\_qd} = (Q_{1\_qd} + Q_{2\_qd})/2)$ are used to ensure proper load-sharing. The reason why the average power method is used in this research is to overcome the sensitivity about voltage and current measurement errors.

Figure 3.12 shows a Simulink block diagram for the real power sharing control. In this figure, phase angle $(\Delta\theta_{phase})$ is chosen as the control variable instead of frequency

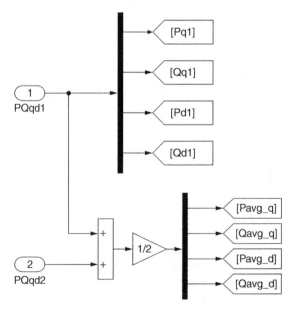

FIGURE 3.11 Simulink model for average power (P_{avg_qd} and Q_{avg_qd}) generation.

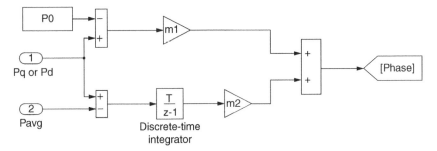

FIGURE 3.12 Real power (P_q or P_d) sharing control.

used as a control variable in a conventional droop method because the phase angle is not varied due to constant real power in steady-state time, and then it makes the frequency remain at nominal value (60 Hz), unlike some frequency deviations of the conventional droop method. As shown in Fig. 3.12, the difference between real-time active power (P_q or P_d) of each unit and the rating real power (P_{0_qd}) of each unit is multiplied by a droop coefficient (m_1). In addition, the difference between the active power (P_q or P_d) of each DGS unit and the average real power (P_{avg_qd}) is finally equal to zero by a discrete-time integrator in steady-state time. As a result, the active power between the DGS units is properly shared according to the power ratings of each DGS unit.

The Simulink block diagram for the reactive power (Q_q or Q_d) sharing control is implemented in Fig. 3.13. As shown in Fig. 3.13, the amplitude of a reference voltage (V_{max_qd}) is decided by the average reactive power (Q_{avg_qd}) and the reactive power of each unit. The difference between real-time reactive power (Q_q or Q_d) of each unit and the rating reactive power (Q_{0_qd}) of each unit is multiplied by a droop gain (n_1). In addition, by a discrete-time integrator, the reactive power (Q_q or Q_d) of each DGS unit is also finally equal to an average reactive power (Q_{avg_qd}) in steady-state time. In order to generate the amplitude of the reference load voltage, the nominal

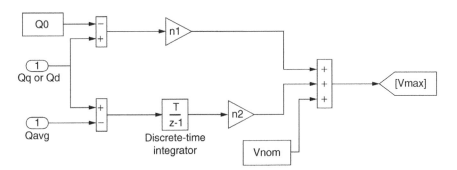

FIGURE 3.13 Reactive power (Q_q or Q_d) sharing control.

load voltage $\left(V_{\text{nom}}\left(480 \cdot \sqrt{2}\right)\right)$ is also added to the results by the droop gains of the average reactive power ($Q_{\text{avg\_}qd}$) and the reactive power (Q_q or Q_d).

From Figs. 3.12 and 3.13, the phase angle and voltage amplitude of the reference load voltage are obtained from the following equations:

Q-Axis

Phase angle:

$$\begin{aligned}
\phi_q(k+1) &= \phi_q(k) + m_2(P_q - P_{\text{avg\_}q}), \\
\Delta\theta_q(k) &= \phi_q(k) + m_1(P_q - P_{0\_q}), \\
\theta_q(k) &= \theta_{ref}(k) + \Delta\theta_q(k).
\end{aligned} \tag{3.39}$$

Amplitude:

$$\begin{aligned}
V_q(k+1) &= V_q(k) + n_2(Q_q - Q_{\text{avg}}), \\
\Delta V_q(k) &= V_q(k) + n_1(Q_q - Q_{0\_q}), \\
V_{\text{max\_}q}(k) &= V_{\text{nom}} + \Delta V_q(k).
\end{aligned} \tag{3.40}$$

Voltage Reference:

$$V_{ref\_q}(k) = V_{\text{max\_}q} \cdot \sin(\theta_q(k)). \tag{3.41}$$

D-Axis

Phase angle:

$$\begin{aligned}
\phi_d(k+1) &= \phi_d(k) + m_2(P_d - P_{\text{avg\_}d}), \\
\Delta\theta_d(k) &= \phi_d(k) + m_1(P_d - P_{0\_d}), \\
\theta_d(k) &= \theta_{ref}(k) + \Delta\theta_d(k),
\end{aligned} \tag{3.42}$$

Amplitude:

$$\begin{aligned}
V_d(k+1) &= V_d(k) + n_2(Q_d - Q_{\text{avg\_}d}), \\
\Delta V_d(k) &= V_d(k) + n_1(Q_d - Q_{0\_d}), \\
V_{\text{max\_}d}(k) &= V_{\text{nom}} + \Delta V_d(k).
\end{aligned} \tag{3.43}$$

Voltage Reference:

$$V_{ref\_d}(k) = V_{\text{max\_}d} \cdot \sin(\theta_d(k)), \tag{3.44}$$

where m_1 and n_1 are droop gain for real power and reactive power, respectively; m_2 and n_2 represent the error gain between the average real power ($P_{\text{avg\_}qd}$) and

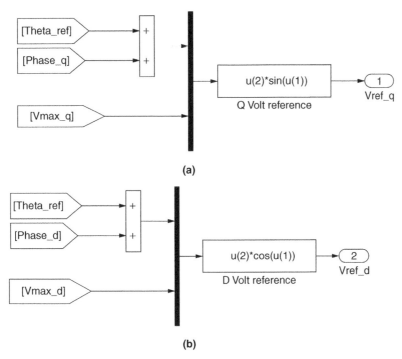

FIGURE 3.14 Reference voltage (V_q and V_d) generation. (a) V_q reference voltage. (b) V_d reference voltage.

real power (P_{qd}) of each unit and the average reactive power (Q_{avg_qd}) and reactive power (Q_{qd}) of each unit, respectively; P_{0_qd} and Q_{0_qd} are the real and reactive power rating of each unit, respectively; and V_{nom} and f_{nom} are the nominal voltage amplitude $\left(480 \cdot \sqrt{2}\ V\right)$ and nominal frequency (60 Hz), respectively (θ_{ref}: $2 \cdot \pi \cdot f_{nom} \cdot t$).

Finally, according to the proposed load-sharing control method, a Simulink block diagram of the reference load voltage (V_{ref_q} and V_{ref_d}) is described by the above equations as shown in Fig. 3.14.

3.6.2 Harmonic Sharing Control Loop

The only above-combined droop method and average power control method does not guarantee the harmonic components of the load current to share because it affects only the phase and magnitude of the fundamental output voltage. A means is required to share the harmonic components of the load currents based on its harmonic contents. The control gains affecting the harmonics in the voltage controller that was mentioned in previous voltage controller section can be adjusted based on the harmonic contents of the load current at that harmonic frequency. For example, the pole frequencies of the harmonic compensator can be shifted by the harmonic

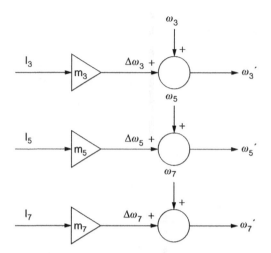

FIGURE 3.15   Harmonic sharing droop control loop.

contents of the load current at those frequencies. This is illustrated in Fig. 3.15, where I_3, I_5, and I_7 denote the harmonic load currents at third, fifth, and seventh harmonics, respectively, and ω_3', ω_5', and ω_7' are the third, fifth, and seventh harmonic frequencies of the harmonic compensator poles. First of all, the harmonic droop loop added has the advantage that it does not degrade the fundamental component; it only affects the individual harmonic when it exists.

The ω_3', ω_5', and ω_7' are computed from the harmonic droop control loop in Fig. 3.15:

$$\omega_3' = \omega_3 + m_3 I_3, \qquad \omega_3 = 2\pi \cdot 3 \cdot 60,$$
$$\omega_5' = \omega_5 + m_5 I_5, \qquad \omega_5 = 2\pi \cdot 5 \cdot 60,$$
$$\omega_7' = \omega_7 + m_7 I_7, \qquad \omega_7 = 2\pi \cdot 7 \cdot 60.$$

So harmonic frequencies in Eqs. (3.32) and (3.33) need to be modified from ω_3, ω_5, and ω_7 to ω_3', ω_5', and ω_7' in order to ensure harmonic load-sharing.

3.7   SIMULATION RESULTS

Figure 3.16 shows configuration of the simulated distributed generation systems. This configuration consists of two DGS units and two loads. In real circuit model, wire impedances (Z_1 and Z_2) and interconnected tie-line impedance are modeled because these significantly affect load-sharing between the DGS units.

To simulate Fig. 3.16 with MATLAB/Simulink, the configuration is modeled as Fig. 3.17. This model is also composed of two DGS units, two loads, wire impedances (Z_1 and Z_2), and tie-line impedance. In particular, power information (PQ_{qd1} and

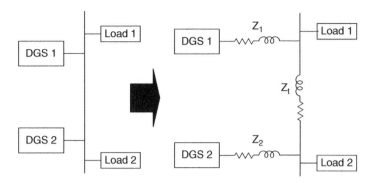

FIGURE 3.16 Configuration of the simulated system.

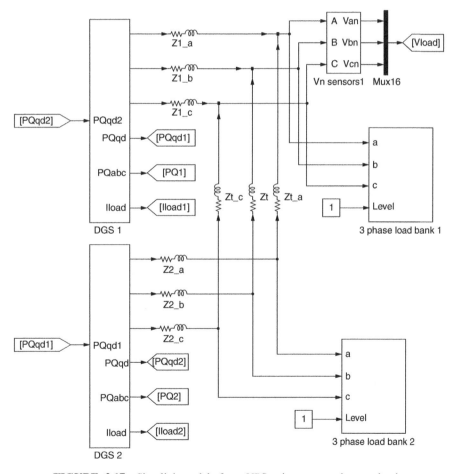

FIGURE 3.17 Simulink model of two UPS units connected to two loads.

PQ_{qd2}) such as the real power (P) and the reactive power (Q) is exchanged between two DGS systems to ensure proper load-sharing.

Simulations were performed using MATLAB/Simulink v6.1 with Power System Block-set (PSB). For speed of simulations, the PWM Bridge IGBT inverter has been modeled as an ideal voltage-controlled source with a delay of half the sampling time of the actual PWM signal. Two linear transformers are used for the isolation transformers, and a series inductance and resistance representing the leakage impedance and losses of each transformer are, respectively, 3% p.u. 0.1% p.u. The series inductance and resistance of the transformers are denoted as L_T and R_T, respectively.

The nominal values of the circuit components for simulations are as follows:

Rated output power of each DGS system = 600 kVA

$V_{DC} = 2200$ V

AC load voltage = 480 V (RMS)

Primary side voltage of Δ/Y transformer = 980 V (RMS, line to line)

Secondary side voltage of Δ/Y transformer = 480 V (RMS, line to neutral)

AC output frequency = 60 Hz

PWM frequency = 5.4 kHz

T_s (sampling period) = 185 μsec

T_d (delay time) = $T_s/2$

$L_f = 5$ mH

$C_f = 300$ μF

$L_T = 31$ μH

$R_T = 0.0115$ Ω

$C_L = 90$ μF

As explained in a previous section, each PWM inverter's output voltage and current are controlled using a dual-loop control system, with the outer loop (RSC) controlling the output voltage and the inner loop (DSMC) controlling the inverter current.

The droop coefficients are given as follows:

$m_1 = -2$e-6 rad/watt

$m_2 = -10$e-6 rad/watt

$n_1 = -2$e-4 V/var

$n_2 = -20$e-4 V/var

$m_3 = m_5 = m_7 = -2$e-2

In this study, the two PWM inverters are assumed to have identical characteristics, that is, they have matched circuit components equal to their nominal values. To verify the performance of the proposed droop method for load-sharing control, tie-line impedance, wire impedances mismatches, and voltage/current sensor measurement error mismatches are considered in this simulation.

The following cases are simulated:

All Cases:

Unit 1: $Z_1 = R_1 + jX_1(R_1 = 0.01\ \Omega/L_1 = 0.5\ \text{mH})$

Voltage measurement error: V_p [−0.1%, +0.1%, −0.1%], V_L [+0.1%, +0.1%, −0.1%]

Current measurement error: I_i [+0.1%, −0.1%, −0.1%], I_L [+0.1%, −0.1%, −0.1%]

Unit 2: $Z_2 = R_2 + jX_2\ (R_2 = 0.02\ \Omega/L_2 = 1\ \text{mH})$

Voltage measurement error: V_p [+0.1%, −0.1%, +0.1%], V_L [−0.1%, −0.1%, +0.1%]

Current measurement error: I_i [−0.1%, +0.1%, +0.1%], I_L [−0.1%, +0.1%, +0.1%]

$Z_t = R_t + jX_t\ (R_t = 0.02\ \Omega/L_t = 1\ \text{mH})$

Case 1:

Power ratings of DGS unit 1 and 2: 600 k VA, respectively

Load 1: $P_{\text{load1}} = 480\ \text{kW}/Q_{\text{load1}} = 360\ \text{kVar}$ (p.f = 0.8)

Load 2: $P_{\text{load2}} = 480\ \text{kW}/Q_{\text{load2}} = 360\ \text{kVar}$ (p.f = 0.8)

Case 2:

Power ratings of UPS unit 1 and 2: 600 k VA, 500 kVA, respectively

Load 1: $P_{\text{load1}} = 400\ \text{kW}/Q_{\text{load1}} = 300\ \text{kVar}$ (p.f = 0.8)

Load 2: $P_{\text{load2}} = 480\ \text{kW}/Q_{\text{load2}} = 360\ \text{kVar}$ (p.f = 0.8)

Case 3:

Power ratings of UPS unit 1 and 2: 600 k VA, respectively

Load 1: $P_{\text{load1}} = 480\ \text{kW}/Q_{\text{load1}} = 360\ \text{kVar}$ (p.f = 0.8)

Load 2: $P_{\text{load2}} = 240\ \text{kW} => 480\ \text{kW}$ (after 1.6 sec)
$\qquad Q_{\text{load2}} = 180\ \text{kVar} => 360\ \text{kVar}$

Case 4:

Power ratings of UPS unit 1 and 2: 600 k VA, respectively

Load 1: Three-phase bridge diode ($C_{\text{DC1}} = 10000\ \mu\text{F}$ and $R_{\text{L1}} = 3.25\ \Omega$)

Load 2: Three-phase bridge diode ($C_{\text{DC2}} = 10000\ \mu\text{F}$ and $R_{\text{L2}} = 3.25\ \Omega$)

All cases are assumed that Z_2 is twice of Z_1, the signs of voltage/current sensor errors of unit 1 are opposite to those of unit 2. In Case 1, and power rating and load of unit 1 are equal to those of unit 2. In Case 2 it is supposed that power ratings and loads of two DGS units are different, unlike Case 1. In Case 3 it is assumed that load 2 becomes twice after 1.6 seconds. Finally, Case 4 is simulated under nonlinear loads with three-phase bridge diode. In this simulation, the first three cases (Figs. 3.18–3.20) were done for linear loads, while the last case was simulated under

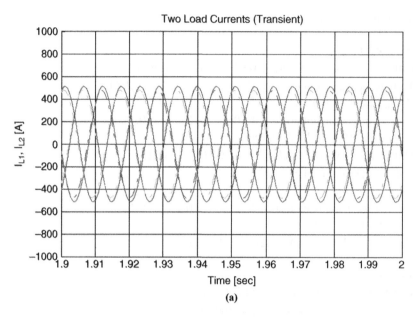

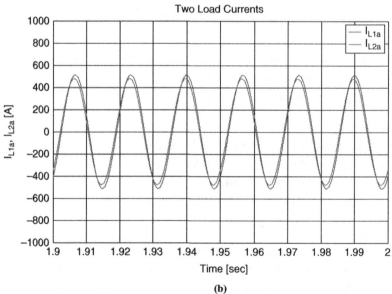

FIGURE 3.18 Simulation results for Case 1. (a) Load currents (V_{L1} and V_{L2}). (b) "A"-phase load currents (V_{L1a} and V_{L2a}). (c) Real powers (P_1 and P_2). (d) Reactive powers (Q_1 and Q_2).

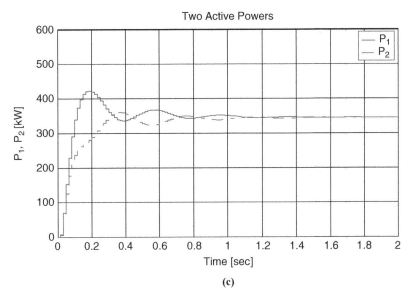

(c)

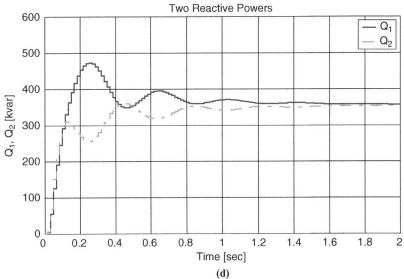

(d)

FIGURE 3.18   (Continued)

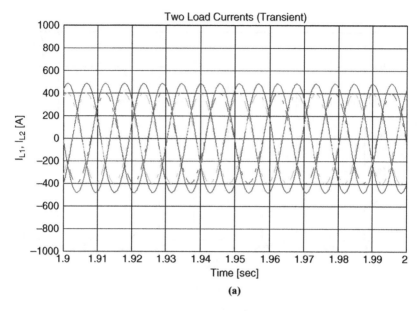

(a)

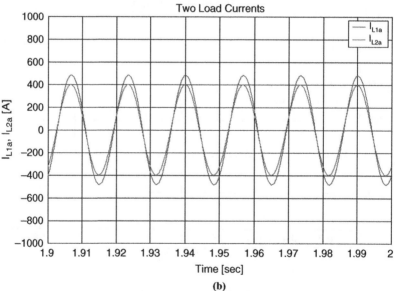

(b)

FIGURE 3.19 Simulation results for Case 2. (**a**) Load currents (V_{L1} and V_{L2}). (**b**) "A"-phase load currents (V_{L1a} and V_{L2a}). (**c**) Real powers (P_1 and P_2). (**d**) Reactive powers (Q_1 and Q_2).

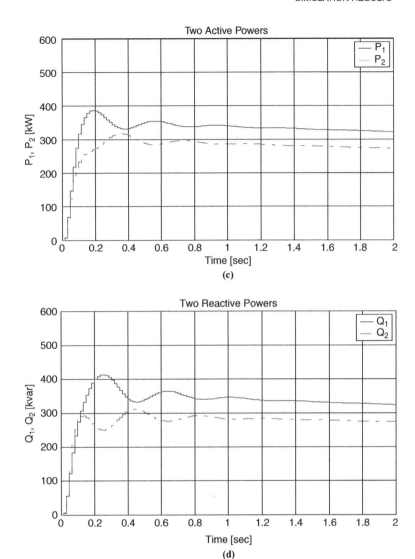

FIGURE 3.19 (Continued).

nonlinear loads. Fig. 3.17 shows very good load-sharing of the real and the reactive powers under the conditions that both units and loads are identical. Fig. 3.19 shows the results under different power ratings and loads, and these results also show that the loads are properly shared according to the power capability of each unit. As shown in Fig. 3.20, even when the load 2 is increased suddenly to twice its value after 1.6 sec, the results definitely show so good a power sharing. Two nonlinear loads that consist of a three-phase bridge, a large capacitor, and a small resistor are implemented, and the results are also given in Fig. 3.21.

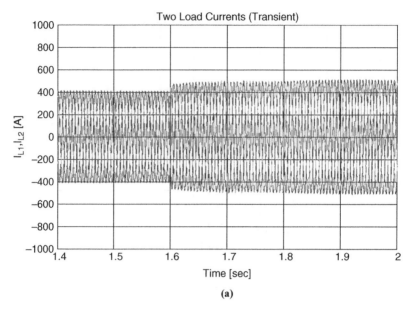

(a)

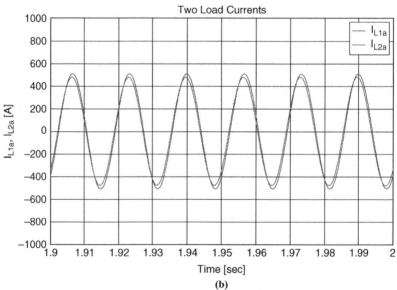

(b)

FIGURE 3.20   Simulation results for Case 3. (**a**) Load currents (V_{L1} and V_{L2}). (**b**) "A"-phase load currents (V_{L1a} and V_{L2a}). (**c**) Real powers (P_1 and P_2). (**d**) Reactive powers (Q_1 and Q_2).

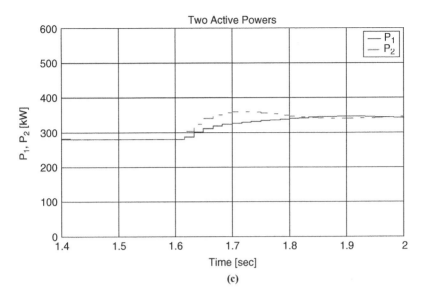

(c)

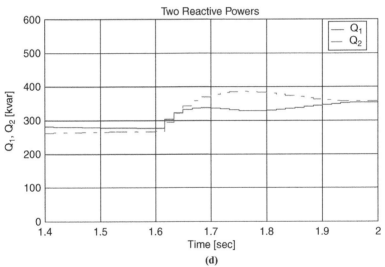

(d)

FIGURE 3.20 (Continued).

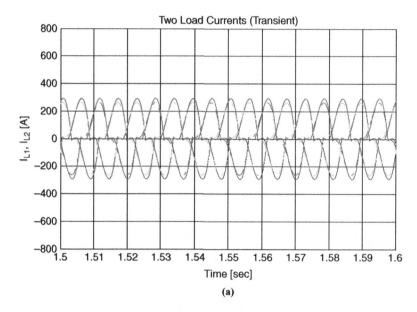

(a)

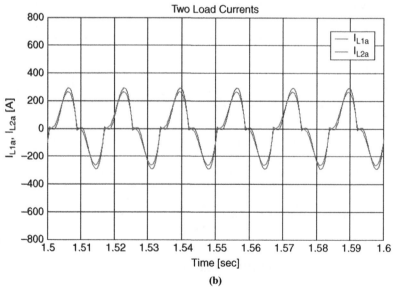

(b)

FIGURE 3.21   Simulation results for Case 4. (**a**) Load currents (V_{L1} and V_{L2}). (**b**) "A"-phase load currents (V_{L1a} and V_{L2a}). (**c**) Real powers (P_1 and P_2). (**d**) Reactive powers (Q_1 and Q_2).

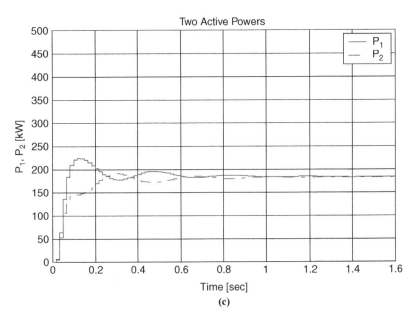

(c)

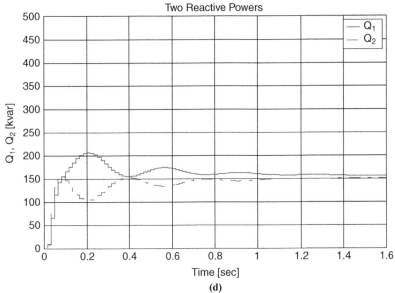

(d)

FIGURE 3.21   (Continued.)

3.8 CONCLUSIONS

This chapter has described a new droop control method that can properly control the load-sharing such as the real, reactive, and harmonic powers for parallel operation of distributed generation systems in a standalone AC power system. The proposed scheme is implemented by a combined droop method, an average power control method, and a harmonic sharing control loop using power information exchange between each DGS unit. In particular, the theory of the average power control method can significantly reduce the sensitivity about voltage and current measurement errors, and the harmonic control loop is also added to guarantee harmonic power sharing under nonlinear loads.

In addition, to support this control algorithm, two feedback controllers are used: a voltage controller using a robust servomechanism controller (RSC) and a current controller using a discrete-time sliding mode controller (DSMC). The RSC with servocompensator and stabilizing compensator is adopted in the outer voltage control loop due to its capability to perform zero steady-state tracking error under unknown load and eliminate harmonics of any specified frequencies with guaranteed system stability. The DSMC is used in the inner current loop to limit the inverter current under overload condition because of the fast and no-overshoot response it provides. In this chapter, two three-phase PWM inverters (600 kVA) operating in parallel are implemented by digital computer simulation using MATLAB/Simulink, and space vector pulse width modulation (SVPWM) is used as a PWM technique. In general, it is well known that wire impedance mismatches and voltage/current sensor measurement error mismatches heavily degrade the load-sharing of paralleled power converters. To verify the power-sharing performance of proposed control strategy, the above critical factors are all considered under all simulation conditions. With reference to the above simulation results (Figs. 3.18–3.21), it is shown that the proposed control method is very effective for each DGS to properly share the loads such as the real, reactive, harmonic powers in a standalone AC power system.

CHAPTER 4

POWER CONVERTER TOPOLOGIES FOR DISTRIBUTED GENERATION SYSTEMS

4.1 INTRODUCTION

In this chapter, we will give a second look at important of distributed generation as an approach in integration of green and renewable energy sources. This will put into focus the importance of inverter topology for distributed generation. Furthermore, we present the concepts that have been reported in the literature. The reference section at the end of this book will give a wealth of information sources for readers. Finally, we will present the important system topologies that we will use in a later chapter of this book.

4.2 DISTRIBUTED GENERATION SYSTEMS

The need for electric energy is never ending. Along with the growth in demand for electric power, sustainable development, environmental issues, and power quality and reliability have become concerns. Electric utilities are becoming more and more stressed since existing transmission and distribution systems are facing their operating constraints with growing load. Under such circumstances, distributed generation (DG) with alternative sources are an urgent task. Distributed generation entails using many small generators of 2 to 50-MW output, situated at numerous strategic points throughout cities and towns, so that each provides power to a small number of consumers nearby and dispersed generation refers to use of even smaller generating

Integration of Green and Renewable Energy in Electric Power Systems. By A. Keyhani, M. N. Marwali, and M. Dai
Copyright © 2010 John Wiley & Sons, Inc.

units, of less than 500-kW output and often sized to serve individual homes or businesses. Later publications tend to combine the two categories into one (i.e., distributed generation), to refer to power generation at customer sites to serve part or all of customer load or as backup power, or, at substations, to reduce peak load demand and defer substation capacity reinforcements.

Distributed generation is not a new concept since traditional diesel generator as a backup power source for critical load has been used for decades. However, due to its low efficiency, high cost, and noise and exhaust, a diesel generator would be objectionable in any applications but emergency and fieldwork, and it has never become a true distributed generation source on today's basis. What endows new meaning to this old concept is technology. Environmental-friendly renewable energy sources (such as photovoltaic devices and wind electric generators), clean and efficient fossil-fuel technologies (such as micro-gas turbines), and hydrogen electric devices (fuel cells) have provided great opportunities for the development of distributed generation. Gas-fired micro-turbines in the 25 to 100-kW range can be mass produced at low cost using airbearing and recuperation to achieve reasonable efficiency at 40% with electricity output only and 90% for electricity and heat micro-cogeneration. Fuel cells have the virtue of zero emission, high efficiency, and reliability and therefore have the potential to truly revolutionize power generation. The hydrogen can be either directly supplied or reformed from natural gas or liquid fuels such as alcohols or gasoline. Individual units range in size from 3 to 250 kW or even larger megawatt size. The fastest growing renewable energy source is wind power. On a worldwide basis, available wind energy exceeds the presently installed capacity of conventional energy sources by a factor of four. Photovoltaic systems can be used in a variety of sizes and show better potentials in those areas with high intensity and reliability of sunlight. Besides these power generators, storage technologies such as batteries, ultracapacitors, and flywheels have also been significantly improved. Flywheel systems can deliver 700 kW for 5 sec while 28-cell ultracapacitors can provide up to 12.5 kW for a few seconds. To apply the above generation and storage technologies in an DG environment involves new technical problems.

DG units require power electronics interfacing and different methods of control and dispatch. A DG unit should be able to operate under either island mode or grid-connected mode. In island mode, it should provide steady, low regulation error, low total harmonic distortion (THD), and fast response AC power under various load disturbances. In grid-connected mode, it should give steady-state decoupled active power P and reactive power Q control and proper behavior under connecting, disconnecting, and reclosing operations. If multiple units are paralleled on the same terminal or bus, correct load sharing should be performed among the units.

A DC/AC voltage source inverter (VSI) is the most widely used interface for DG units, which involves many topology and control aspects under different operating conditions. Only with satisfactory control performance of each individual unit can paralleling two or more inverters or connecting one or more inverters to the power system be conducted. This involves P and Q control under various local load characteristics and operating conditions. As stated above, the tremendous complexity in the power electronics interfaces for DG units creates many research problems as well as many possibilities to advance technologies. Many of the problems have been solved or partly solved, while many are still left unsolved or even unfound. In general, a practically functioning DG system has to properly solve possible technical problems in the following three categories: control of a single inverter unit with quality voltage output in island mode, control of line real and reactive power flowing between a DG unit and the utility grid in grid-connected mode, and control of front-end power generation or conversion for high performance and low overhead. Due to the great potential of DG technologies, these research problems deserve special attentions and warrant careful further investigations.

In this book, the above problems will be addressed by presenting the following information - problem descriptions, proposed solutions, related analysis, simulation and experimental results, and conclusions and discussions. a literature review will be given within the scope of research as mentioned above about DG control technologies and the existing solutions will be evaluated. Specific research problems will be described.

There are a number of supporting concepts and technologies that must be described. These topics are:

1. Voltage and current control of individual inverters in island mode.
2. The system topology.
3. Robust stability issues.
4. Pulse width modulation techniques.
5. Line-interactive operation of inverters and control of P and Q.
6. Front-end rectifier control in controlled AC–DC–AC systems.

4.3 VOLTAGE AND CURRENT CONTROL OF INDIVIDUAL INVERTERS IN ISLAND MODE

In island mode, an inverter unit needs to control load voltage as it is supplying local load with quality power. There are a number of control methods that are used to control the inverter voltage. These methods are: PID control, model-based linear control, robust control, sliding mode control, and internal-model-principle-based control.

Conventional PI Control Method. This is a classical method of voltage control of an inverter based on proportional-integral (PI) regulation under stationary reference frame where the PI regulators have to track sinusoidally varying inputs. Since a PI controller only guarantees zero steady-state error under DC reference input, this control technique cannot be convincing in control performance. This method has been extended by using proportional control plus model-based compensation. It does not utilize the information that the reference input is a 60-Hz sine wave, so that the control design has to be able to handle arbitrary input, which is unlikely to yield a good control performance for DG applications. Another variation in this method is designed based on a dual loop proportional control scheme for single-phase half-bridge inverters in island mode.

State Feedback-Based Controls. Standard linear control theory has been employed to control the inverter voltage. One approach is based an analog control algorithm using a canonical lead-lag compensator based on a transfer function model. The extension of this method is a state-space-model-based state feedback control technique. This method has also used a linear quadratic index on top of traditional linear state feedback control to achieve optimized performance.

Robust Control. $H\infty$ design procedure can be used for a inverter control to improve robust stability under model uncertainty and load disturbance. However, the control performance under nonlinear load is not satisfactory.

Sliding Mode Controls. Sliding mode control has also been used in inverter control due to its robustness and overshoot-free fast tracking capability. This technique is used with discontinuous control and discrete control and implementing it with a digital controller. In this technique, the control variable in each sampling period is calculated based on the plant model and feedback quantities. The control is continuous and the chattering problem does not exist. The results show good performances under both linear and nonlinear loads. This method has been extended using deadbeat current control and proportional voltage control. The deadbeat control concept is the same as the discrete-time sliding mode control when the plant model parameters are known.

Internal Model Principle and Reference Frames. This method states that asymptotic tracking of controlled variables toward the corresponding references in the presence of disturbances (zero steady-state tracking error) can be achieved if the models that generate these references and disturbances are included in the stable closed-loop systems. Actually a PI controller is an example of using the internal model principle in that the integral term models the mode of a step input and therefore results in zero steady-state error tracking DC reference input. However, this is no longer true if the reference signal is an AC quantity. In three-phase systems, reference frame transformation from a stationary *ABC* frame to a synchronous rotating *dq* reference frame can transform AC quantities in synchronous frequency into DC quantities that can be handled by a PI controller. It has to be noted that a synchronous reference frame can only transform components in the synchronous frequency into DC while all other frequency components are still AC. In three-phase DG systems, if the fundamental frequency components are transformed into a synchronous reference frame, all harmonic components are still AC quantities in the same synchronous reference frame. This method has been used as a proportional control scheme for a three-phase inverter based on small-signal analysis in a synchronous reference frame only for the fundamental frequency. The method has been utilized as a nonlinear prediction technique to handle harmonic components in a fundamental synchronous reference frame. Due to the limitation of a single fundamental synchronous reference frame in handling harmonics, ideas have been raised for having multiple rotating reference frames corresponding to multiple frequency components including the fundamental and harmonics as well. This method has been used as a three-phase inverter controller with multiple synchronous reference frames. This technique requires information of magnitudes and phase angles of each frequency components from phase-locked loops (PLL), which increases the complexity of the solution. The proposed control technique is also based on multiple rotating reference frames trying to convert multiple frequency components into DC. This technique needs gain and phase correction in each reference frame, which also increases the overall complexity of the solution. In general, the multi-rotating frame type of technique provides a systematical solution to achieve zero steady-state error for multiple frequency components while the trade-off is the high complexity. The internal model principle can be better used in a different way where the modes of all frequencies of interest are modeled in the same reference frame so that the steady-state tracking errors of all modeled frequencies can reach zero. Typically, a stationary reference frame for three-phase systems is used where all AC frequency components remain AC since there is no necessity to take advantage of DC quantities given the capability of handling AC directly using the internal models. Clarke's transformation from ABC stationary reference frame to $\alpha\beta0$ stationary reference frame provides decoupling between the axes and enables independent modeling and control in each dimension and hence it is widely used. There is no reference frame issue at all in single-phase systems since all original quantities are in stationary reference frame inherently. Repetitive control is a specific implementation of internal model principle in a single reference frame, which eliminates periodical tracking error or disturbance whose frequency is less than half the sampling frequency. For a single-phase half-bridge inverter, a standard single-loop repetitive controller can be designed and implemented in discrete time, which yields acceptable steady-state performance but slow transient response. The method can be extended by using a state feedback plus integral controller that has been added to provide better response to instantaneous disturbances, and the idea has been proved effective by the presented results. Different from repetitive control, the internal model principle can be used only to eliminate periodical tracking error and disturbance with specified frequency, which is generally enough for inverter control for DG applications since THD is nearly caused by low-order harmonics. All works in this area take advantage of the concept of *generalized integrator* which expands the functionality of the integral term of a PI controller for multiple frequency components in the same reference frame. The method is used with servocontroller and linear quadratic optimization in their control approach for a three-phase inverter. This approach is based on the robust servomechanism control theory that yields a THD of 2.7% under nonlinear load with a crest factor of 3:1, which is satisfactory.

4.4 THE SYSTEM TOPOLOGY

Three-phase inverters can be designed to have a three-phase three-wire system where the inverter itself does not provide a neutral point. Typically, a Δ/Yg transformer is used with the secondary center grounded before the inverter powering the load or being connected to utility grid as shown in Fig. 4.1. In this topology, the three-wire system on the Δ side only has two independent dimensions and 0-axis current cannot flow, which makes the system relatively easy to control. The drawback is the existence of the costly, heavy, and bulky transformer.

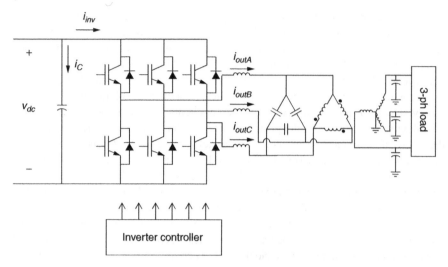

FIGURE 4.1   The three-phase three-wire inverter topology.

Multiple grounding topologies for inverter systems, where a three-phase four-wire inverter topology with a grounded center of the DC bus is mentioned, are shown in Fig. 4.2. The benefit of this topology is that the undesirable transformer can be removed.

The topology for three independent single-phase half-bridge inverters is presented in Fig. 4.3. The topology of three-phase four-wire transformerless inverter topology with the three-phase four-leg inverter is shown in Fig. 4.4. Unlike the split DC bus

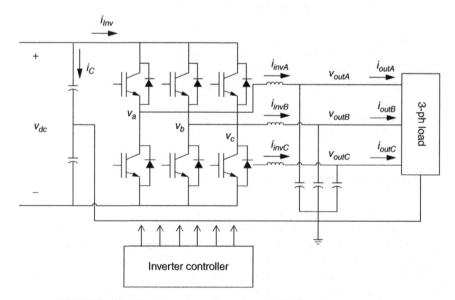

FIGURE 4.2   The three-phase four-wire split DC bus inverter topology.

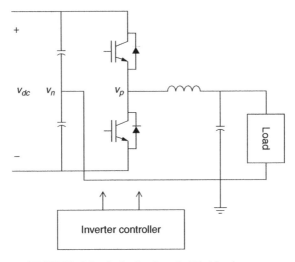

FIGURE 4.3   A single-phase half-bridge inverter.

topology mentioned above, the four-leg inverter uses a fourth leg whose mid-point is used as the neutral point of the inverter. This topology is equivalent to combining three full-bridge single-phase inverters together as shown in Fig. 4.5 with a shared neutral leg.

4.4.1   Robust Stability

A feedback control system is said to achieve robust stability if it remains stable for all considered perturbations in the plant. In feedback-controlled PWM inverter

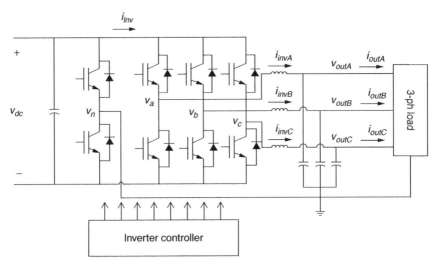

FIGURE 4.4   The three-phase four-leg inverter topology.

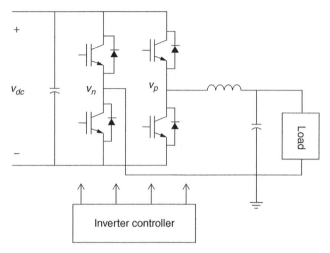

FIGURE 4.5 A single-phase full-bridge inverter.

systems (e.g., an inverter-based three-phase distributed generation unit operated in island mode), load disturbance, noise, and parametric uncertainty of the electrical components in the circuit are the major plant perturbations that have significant impacts on both system stability and performance.

4.4.2 Pulse Width Modulation Techniques

Pulse width modulation (PWM) is an essential but not only technique to control a switched-mode power converter using self-commutated devices, including the DC/AC inverter and AC/DC rectifier mentioned above. Alternative modulation techniques include the hysteresis technique and delta modulation where the switching frequency is not a constant, which tends to cause harmonic distortions in its output waveform and therefore limits the application of such techniques. A number of PWM techniques have been developed and used in power converter controls to generate sinusoidal waveforms. However, only three of them have become standards and have most often been applied into practice: naturally sampled sine PWM (NSPWM), uniformly or regularly sampled sine PWM (USPWM), and space vector PWM (SVPWM). NSPWM uses a modulation signal (i.e., the sine wave) to be compared to a high-frequency carrier (i.e., a triangle waveform); and the compared result, which is a logic signal, is used to determine the ON or OFF state of the power switches. NSPWM does not control the position of the pulses it generates in each cycle, and the minimum pulse width is not controlled. USPWM still uses a triangle carrier signal at switching frequency, but it only uses the comparison result to determine the ON or OFF duration of the switches but not the pulse position. The pulse position is uniformly controlled—for example, put at the center of each switching cycle. SVPWM maps eight switching patterns of a three-phase full-bridge converter into six 60°-apart space vectors on the same plane and two 0-axis vectors perpendicular to the plane, and a reference vector on the plane is used as a modulation signal and determines the time average of these switching patterns in each PWM cycle. If the reference vector rotates on the plane from sector to sector, a sine wave is modulated in the pulses.

All inverter control techniques listed above are based on an assumption that the DC bus voltage is fixed and well-regulated or that its change does not affect the control on the load side. However, this assumption is not always true in a DG environment where fuel cells, wind generators, or photovoltaic modules could be the source with an unregulated or poorly regulated DC voltage. Even though in most cases, a properly developed PWM scheme can compensate for the DC bus voltage change inherently so that the load side cannot encounter any effect of the DC voltage change, this change may still undermine the performance of PWM inverter under some situations.

4.4.3 Line-Interactive Operation of Inverters and Control of *P* and *Q*

Line-interactive inverter systems with a local load need active filtering capability of the inverter to compensate for the effect of harmonic corrupted load current by pumping compensating current into the power system through real-time control. In this type of application, the goal is to make the line current as sinusoidal as possible. The inverter control technique for photovoltaic systems connected with utility grid can use standard space vector PWM with only a linear local load. A special topology, referred to as series–parallel topology, utilizes power dragged from a utility grid through a rectifier to condition the utility line current through an inverter. The rectifier is connected to the utility grid through a transformer in series with the power line, and the inverter is paralleled with the power line; this is how the topology is named.

4.4.4 Current Quality of Line-Interactive Inverters

When an inverter is connected to a power system, the terminal voltage is governed by the power system but the current waveform is still controllable. Although current waveform control is one of the goals of the line-interactive inverter control, power control (including *P* and *Q* controls) is the eventual objective to be achieved by a power electronics interface in DG environment. DG systems may have significant impact on power system stability if not properly compensated in reactive power. Also, DG systems can have significant impacts on transmission system stability at heavy penetration levels, where penetration is defined as the percentage of DG power in total load power in the system. A DG unit affects the system stability by generating or consuming active and reactive power. Therefore, the control of power on DG system determines its impact on its local the utility grid. If the power control performs well, the DG unit can be used as means to enhance the system stability and improve power quality; otherwise it could undermine the system stability.

A line-interactive uninterruptible power supply (UPS) is able to pick up the load at power system failure and reverse power flow direction to battery charging when the power line is restored However, the power flow control in line-interactive UPS does not match the requirement of DG systems by far. A three-phase AC–DC–AC power conversion system can also be used for interfacing small wind turbines to utility grid. This system has been developed for both island and grid-connected operations.

The power control concept for synchronous generator paralleled with power system into the application of grid connected inverters, that active power P can be controlled by adjusting phase angle of output voltage and reactive power Q, can be controlled by adjusting the magnitude of the output voltage. This can be accomplished by an inverter control technique for line-interactive operation where P and Q can be separately controlled through closed-loop control.

A power control method can also be used for a grid-connected voltage source inverter that achieves good P and Q decoupling and fast power response. However, this approach requires an interface inductor to be connected between the DG output terminal and power system, whose inductance value is assumed known. The existence of the interface inductor creates a higher voltage magnitude change at the DG terminal. This higher voltage at DG terminal will facilitate the regulation of active power, P and reactive power Q. A different power control strategy based on frequency and voltage drop characteristics of power transmission, which allows decoupling of P and Q at steady state, can also be implemented. In this method, power regulation errors ΔP and ΔQ are used to generate output voltage phase angle and magnitude changes, respectively, which decouples P and Q controls in steady state.

4.5 NEWTON–RAPHSON METHOD

The Newton–Raphson Method is an iterative root-finding algorithm that uses the first few terms of the Taylor series of a function $f(x)$ in the vicinity of a expected solution. The Newton–Raphson Method is widely used in solving power flow problems due to its fast convergence property. This technique has also been used in line-interactive power converter systems. The Newton–Raphson power flow analysis is performed in a power system involving AC–DC–AC switch mode power converters. The Newton–Raphson Method is also used to solve for voltage magnitude and phase angle of a unified power flow controller (UPFC).

4.5.1 Front-End Rectifier Control in Controlled AC–DC–AC Systems

If we look one step back toward the feeder of the inverter DC bus, we can find that significant portions of the feeders are controlled by AC/DC rectifiers. In a DG environment, the input AC source can be any gas- or wind-turbine-driven generators or other AC and/or DC systems. No matter what source is used, a balanced 3-ph input current with low THD is desired, which is called a power factor correction (PFC). There are many existing PFC rectifier topologies. For high-power applications, especially when high performance is required, a continuous conducting mode (CCM) boost rectifier is usually used due to its high efficiency, good current quality, and low EMI emissions. A standard CCM boost rectifier has a full-bridge topology, exactly identical to a three-phase full-bridge inverter. This type of rectifier is controlled by an outer DC voltage control loop and an inner input current control loop, where the voltage regulation error is used to generate the input current command for the inner loop. When the DC voltage is boosted (i.e., greater than the input line voltage amplitude), this rectifier yields excellent control performance. The CCM boost rectifier is also called the PWM rectifier, the boost rectifier, or the controlled rectifier in the literature.

In three-phase four-wire systems, a split DC bus inverter topology can maximize its performance with a three-level controlled rectifier regulating both the top and bottom half voltages of the DC bus. A full-bridge CCM boost rectifier as discussed above can serve the purpose with a modified voltage regulation scheme based on the standard approach described above. Besides the full-bridge topology, VIENNA rectifier topologies can also be used. A VIENNA rectifier is a three-phase three-level rectifier based on a traditional uncontrolled diode rectifier with additional input inductors and six active power switches to achieve the neutral point voltage control. A drawback of the VIENNA rectifiers compared to the full-bridge topology is that it does not allow bidirectional power flow.

In three-phase three-wire systems, if a full-bridge controlled rectifier is used together with an inverter, the impact from the inverter side needs to be taken care of for better control of the rectifier. One typical impact frequently seen is an unbalanced load on the inverter side. It should be noted that the DC voltage ripple problem is caused by either unbalanced inverter load current or unbalanced input voltage supply. However, their control goal was to minimize the DC link voltage ripple instead of improving the input power quality. One can focus on improving instantaneous power balance between the input and output of a rectifier–inverter system and minimizing the DC coupling capacitance to reduce the cost. The lower the DC coupling capacitance, the better the instantaneous power balance the system could yield. However, this is only desirable under balanced load. Once the inverter load is unbalanced, it is apparent that the steady-state inverter output power is no longer a constant, and neither is the inverter input DC power.

A switching function concept for power converters can be used to show the existence of harmonics in DC bus voltage. However, none of these works quantified the harmonic components analytically and used the result to analyze the ripple problem mentioned above.

The control-related issues are major aspects of single-unit operation of switching mode utility interface for DG systems. Single-unit voltage and current control is the basis for DG unit operations in either island mode or grid-connected mode. Many control theories have been applied in this area, and it has turned out that sliding mode control and internal-model-principle-based controls yield better performance under nonlinear load disturbances. Although multiple rotating reference frame techniques can handle the harmonic load disturbances, the stationary-reference-frame-based techniques yield the same performance while cost less overhead. Although the four-leg inverter topology in a three-phase four-wire system performs better than a split DC bus topology in some aspects, the latter requires easier control and uses less power switches and therefore remains an option.

The robust stability of an inverter control technique is an important issue in practice, given parametric uncertainties and load disturbances. Structured singular value μ-based analysis provides evaluation of closed-loop robust stability and the stability margin information and hence can be used as a guideline for control gain tuning.

Uniformly sampled PWM can be made identical to space vector PWM with 0-axis signal injection. Space vector PWM can perform 0-axis control if magnitudes of the two 0-axis vectors are made different. This allows SVPWM to be used in three-phase four-wire systems, but such application has not been seen in the literature.

Power control of a DG unit is necessary under grid-connected running mode. Experimental results about line-interactive inverters are only seen for active filtering purpose or united power flow controller (UPFC) topologies in the literature, while those for power control of DG with local load have not been seen in publications. The Newton–Raphson Method is known as a good nonlinear equation-solving tool and is widely used in power flow problems, including switching mode power converter involved problems, all of which are performed off-line.

As far as the front-end source of a DG unit is concerned, a controlled rectifier should not only provide well-regulated DC voltage but also perform PFC and take balanced input current. In a three-phase three-wire system, unbalanced inverter load may introduce ripple on the DC bus and cause unbalanced input current problem for the rectifier, but no further solution has been reported.

The switch-mode inverter-based DG interface is of interest in (a) island mode, the voltage control problem of a DG inverter with three-phase four-wire transformerless topology for quality power supply to the local load; (b) grid-connected mode, the real and reactive power flow control problem in existence of local load; and (c) a three-phase three-wire AC–DC–AC system, the front-end PFC rectifier control problem with unbalanced inverter load. In this book, all of the above problems will be addressed by proposing a series of new solutions with detailed analysis, simulations, and experimental results.

4.5.2   Voltage and Current Control of A Three-Phase Four-Wire-DG Unit in Island Mode

A three-phase three-leg inverter with split DC bus is one topology to implement three-phase four-wire system with a neutral point seen by the load. Compared to a three-phase three-wire system, it does not have the isolation transformer and provides three-dimensional control. Compared to a three-phase four-leg topology, it saves two power switches and reduces control complexity. Therefore the control problem of the three-phase three-leg inverter with split DC bus is of interest in DG applications. Although a three-phase three-leg inverter with split DC bus topology is a combination of three half-bridge single-phase inverters and control techniques designed for single-phase inverters still work in the three-phase systems, new control problems emerge after the three phases are combined together in that the reference frame issue and the PWM issue become problems. In this report, control to be performed in synchronous

$\alpha\beta0$ reference frame is suggested together with a new modified space vector PWM scheme. Besides, in this book the common control problems shared by both the three-phase system and the single-phase half-bridge topology will also be addressed by presenting a new control solution with detailed analysis of its performances and robust stability.

4.5.3   Power Control of DG in Grid-Connected Mode

Power control, including real power P and reactive power Q controls, of a DG inverter in grid-connected mode with existence of local load is of interest. The challenges come from the fact that the system should also be able to supply quality power to the local load in island mode. Based on this fact, control solutions yielding stability, fast transient response, and less coupling between P and Q are desired.

It is essential to recognize knowledge about how the utility grid helps the control of a DG unit in transients. There has not been any published work addressing the methodology of obtaining the knowledge of the grid and applying the knowledge in DG control in real time. In this book, a power system parameter identification technique and a feed-forward control technique applying the system identification results for real-time implementation will be presented.

If the local load of the DG unit is nonlinear (e.g., diode rectifier sort of load), it tends to draw a harmonic current for the feeder. In island mode, the DG unit is the only feeder. However, in grid-connected mode, how the harmonic current is shared by the DG unit and grid becomes a concern. Harmonic free line current is always desired, and how to let the DG unit take all the harmonic current is an important problem. We will adress the power control technique later.

In practice, the voltage of a utility grid often has harmonic distortion. Whether the DG unit can identify the harmonic components in the grid voltage and compensate for them to maintain clean sinusoidal line current becomes a challenge.

In most cases, from the power control point of view, the inverter topology does not matter. In this book, a three-phase three-wire inverter with an output isolation transformer will be used as the DG unit interfacing with the utility grid because it allows two-dimensional control, which is simpler than three-dimensional control in a four-wire system. The limitation of the two-dimensional topology is that it does not provide zero sequence control and hence cannot maintain balanced three-phase line currents given unbalanced grid voltages.

4.5.4   Front-End Rectifier Control in Three-Phase Three-Wire AC–DC–AC Systems

If a front-end PFC rectifier exists in a three-phase three-wire AC–DC–AC double conversion system, once the inverter load is not balanced, the output power is no longer a constant, which leads to fluctuation of the DC link voltage. On the rectifier side, the ripple corrupted DC link voltage is fed back to the voltage regulator, which generates a fluctuating d-axis current command under a constant DC voltage reference.

If the current regulator of d-axis has high bandwidth, it yields fast current tracking and consequently a fluctuating rectifier output current which causes unbalanced front-end input current in the input current. This situation is undesirable regardless of whether the front end is fed by a power system or a single generator. In this book, the effects of unbalanced inverter load on the DC bus will be analyzed and evaluated. A rectifier control methodology with a method solving this problem will be needed.

CHAPTER 5

VOLTAGE AND CURRENT CONTROL OF A THREE-PHASE FOUR-WIRE DISTRIBUTED GENERATION (DG) INVERTER IN ISLAND MODE

In this chapter we will develop a mathematical model for a three-phase four-wire inverter model. The model will be used in a standalone distributed generation (DG) system. As a first step, MATLAB simulation testbed will be constructed. The control method presented will be used on the simulation testbed. Finally, we will present the experimental testbed. The experimental testbed will be used in conjunction with control technology.

5.1 THE PLANT MODELING

We will use the three-phase four-wire DG unit topology shown in Fig. 5.1. In this circuit, the DC bus voltage is assumed to be an ideal DC voltage source, which can be implemented by a DC-voltage-regulated front end supplied by any distributed sources, including fuel cells, photovoltaic devices, wind turbines, gas turbines, and so on. The control issues of the inverter will be discussed in this chapter.

The inverter has a three-phase three-leg plus split DC bus topology. A full-bridge front-end PFC boost rectifier is assumed to provide regulated DC bus voltage evenly split on the top and bottom halves. The inverter outputs are connected to a three-phase second-order L-C filter, and the filter output voltage is supplied to the load. The control goal of a DG unit in island mode is low steady-state voltage tracking error, low total harmonic distortion (THD) in the output voltage waveforms, and fast transient response to load disturbances under various types

Integration of Green and Renewable Energy in Electric Power Systems. By A. Keyhani, M. N. Marwali, and M. Dai
Copyright © 2010 John Wiley & Sons, Inc.

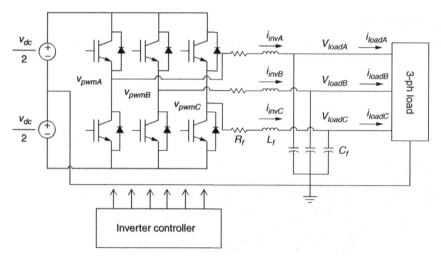

FIGURE 5.1 The three-phase four-wire inverter with a split DC bus.

of load. A new control technique achieving this goal for the DG unit topology will be presented.

5.2 THE BASIC MATHEMATICAL MODEL

The circuit defined in Fig. 5.1 uses the following quantities to describe its behavior. The three-phase inverter output line-to-neutral PWM voltages are v_{pwmA}, v_{pwmB}, and v_{pwmC} and can be represented by vector

$$V_{pwm} = [v_{pwmA} \quad v_{pwmB} \quad v_{pwmC}]^T.$$

The three-phase inverter output currents, which also flow through the filter inductor, are i_{invA}, i_{invB}, and i_{invC} and can be represented by vector

$$I_{inv} = [i_{invA} \quad i_{invB} \quad i_{invC}]^T.$$

The three-phase load voltages, v_{loadA}, v_{loadB}, and v_{loadC}, are the same as the filter capacitor voltages and can be represented by vector

$$V_{load} = [v_{loadA} \quad v_{loadB} \quad v_{loadC}]^T.$$

The three-phase load currents, i_{loadA}, i_{loadB}, and i_{loadC}, can be represented by vector

$$I_{load} = [i_{loadA} \quad i_{loadB} \quad i_{loadC}]^T.$$

Given the filter series resistance R_f, inductance L_f, and capacitance C_f, the system is governed by the following equations:

$$\dot{V}_{load} = \frac{1}{C_f}I_{inv} - \frac{1}{C_f}I_{load}, \tag{5.1}$$

$$\dot{I}_{inv} = \frac{1}{L_f}V_{pwm} - \frac{R_f}{L_f}I_{inv} - \frac{1}{L_f}V_{load}. \tag{5.2}$$

5.3 TRANSFORM THE MODEL INTO STATIONARY REFERENCE FRAME

The control will be performed in Clarke's stationary $\alpha\beta0$ reference frame defined in Fig. 5.2, where the 0-axis is orthogonal to the paper plane, and

$$f_{\alpha\beta0} = \frac{2}{3}\begin{bmatrix} 1 & -\frac{1}{2} & -\frac{1}{2} \\ 0 & \frac{\sqrt{3}}{2} & -\frac{\sqrt{3}}{2} \\ \frac{1}{2} & \frac{1}{2} & \frac{1}{2} \end{bmatrix} f_{ABC} \overset{\Delta}{=} T_{\alpha\beta0}f_{ABC}, \tag{5.3}$$

where f_{ABC} is any vector defined in an ABC reference frame, $f_{\alpha\beta0}$ is its counterpart in an $\alpha\beta0$ reference frame, and the matrix $T_{\alpha\beta0}$ is defined as the transformation matrix from ABC to $\alpha\beta0$.

Apply this reference frame transformation to Eqs. (5.1) and (5.2), it can be obtained that

$$\dot{V}_{load,\alpha\beta0} = \frac{1}{C_f}I_{inv,\alpha\beta0} - \frac{1}{C_f}I_{load,\alpha\beta0} \tag{5.4}$$

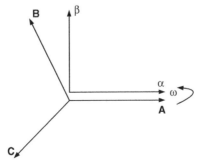

FIGURE 5.2   Clarke's stationary $\alpha\beta0$ to stationary ABC reference frame.

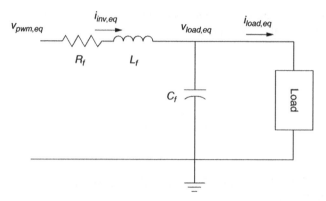

FIGURE 5.3 The one-dimensional equivalent model of the inverter system.

and

$$\dot{i}_{inv,\alpha\beta0} = \frac{1}{L_f}V_{pwm,\alpha\beta0} - \frac{R_f}{L_f}I_{inv,\alpha\beta0} - \frac{1}{L_f}V_{load,\alpha\beta0}. \tag{5.5}$$

It can be observed from Eqs. (2.4) and (2.5) that all dimensions (i.e., α, β, and 0), share the identical dynamics and there is no coupling between any two of the three dimensions. Therefore, without losing generality, the three-dimensional system can be reduced into an equivalent one-dimensional system for convenience in further analysis and control strategy development. In the reduced one-dimensional equivalent system as shown in Fig. 5.3, the load voltage is denoted by $v_{load,eq}$, the inverter current is denoted by $i_{inv,eq}$, the load current is denoted by $i_{load,eq}$, and the inverter PWM voltage can be denoted by $v_{pwm,eq}$. The system equations become

$$\dot{v}_{load,eq} = \frac{1}{C_f}i_{inv,eq} - \frac{1}{C_f}i_{load,eq}, \tag{5.6}$$

and

$$\dot{i}_{inv,eq} = \frac{1}{L_f}v_{pwm,eq} - \frac{R_f}{L_f}i_{inv,eq} - \frac{1}{L_f}v_{load,eq}. \tag{5.7}$$

It can be observed from Fig. 5.3 that this one-dimensional equivalent circuit is exactly the same as one phase of the original three-phase model. Here comes a question about why it is necessary to do the stationary reference frame transformation. This question can be answered as follows. The stationary reference frame transformation is used to transform three-phase quantities from original *ABC* reference frame into Clarke's $\alpha\beta0$ reference frame, and all following control strategy will be developed in this reference frame. It is true that there is no difference in the circuit parameters between the one-dimensional equivalent model and one phase of the three-phase original circuit. However, the system dynamics is described using $\alpha\beta0$ quantities rather than ABC quantities in the one-dimensional equivalent model compared to the case of one phase in the three-phase original circuit. More detailed discussion about the differences of conducting control in these two reference frames and the advantage of the stationary $\alpha\beta0$ reference frame will be presented in later sections.

5.4 CONVERT TO PER-UNIT SYSTEM

For the conveniences of implementation of control algorithms in fixed-point microprocessors and analyzing the control performance, it is beneficial to convert the system model into a per-unit system where all variables and parameters are normalized. Given system rated apparent power S_{rated} and rated output line-to-neutral RMS voltage V_{rated}, the base values can be derived as follows:

$$S_b = \frac{1}{3}S_{rated},$$

$$V_b = \sqrt{2}V_{rated},$$

$$I_b = \sqrt{2}\frac{S_b}{V_{rated}},$$

and

$$Z_b = \frac{V_b}{I_b}.$$

Based on the base values and with v_{load} representing the per-unit load voltage, i_{inv} representing the per-unit inverter current, i_{load} representing the per-unit load current, and v_{pwm} representing the per-unit inverter PWM voltage, the per-unit variables can be obtained as follows:

$$v_{load} = \frac{v_{load,eq}}{V_b},$$

$$i_{inv} = \frac{i_{inv,eq}}{I_b},$$

$$i_{load} = \frac{i_{load,eq}}{I_b},$$

and

$$v_{pwm} = \frac{v_{pwm,eq}}{V_b}.$$

The circuit parameters should also be converted into the per-unit system. Let R represent the per-unit filter resistance, let L represent the per-unit filter inductance,

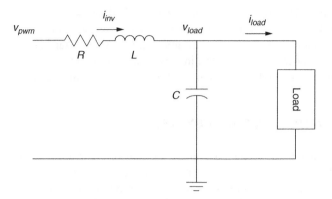

FIGURE 5.4   The per-unit one-dimensional equivalent model of the inverter system.

and let C represent the per-unit filter capacitance. The per-unit parameters can be derived as follows:

$$R = \frac{R_f}{Z_b},$$

$$L = \frac{\omega_1 L_f}{Z_b} \cdot \frac{1}{\omega_1} \cdot \frac{L_f}{Z_b},$$

and

$$C = \frac{1}{\omega_1} \cdot \frac{1}{\dfrac{1}{\dfrac{\omega_1 C_f}{Z_b}}} = C_f Z_b,$$

where $\omega_1 = 2\pi f 1 = 2\pi \times 60 = 120\pi$ (rad/sec) is the fundamental angular frequency. The per-unit one-dimensional equivalent model is depicted in Fig. 5.4 and the system equations become

$$\dot{v}_{load} = \frac{1}{C} i_{inv} - \frac{1}{C} i_{load}, \tag{5.8}$$

and

$$\dot{i}_{inv} = \frac{1}{L} v_{pwm} - \frac{R}{L} i_{inv} - \frac{1}{L} v_{load}. \tag{5.9}$$

The above equations can be rewritten in state-space format as follows:

$$\dot{X} = AX + Bu + Ed, \tag{5.10}$$

where we have the state variable $X = [v_{load} \quad i_{inv}]^T$, the control input $u = vpwm$, the disturbance input $d = i_{load}$, and the coefficients

$$A = \begin{bmatrix} 0 & \dfrac{1}{C} \\ -\dfrac{1}{L} & -\dfrac{R}{L} \end{bmatrix}, \quad B = \begin{bmatrix} 0 \\ \dfrac{1}{L} \end{bmatrix},$$

and

$$E = \begin{bmatrix} \dfrac{-1}{C} \\ 0 \end{bmatrix}.$$

5.5 CONTROL SYSTEM DEVELOPMENT

Given the above per-unit plant model, a dual-loop control structure is used. The inner loop is for inverter current control. A discrete-time sliding mode controller is applied. The outer loop is for the load voltage control. A discrete-time robust servomechanism controller (RSC) is used as the voltage controller. The entire closed-loop system can be illustrated by Fig. 5.5.

In Fig. 5.5, RSC is the robust servomechanism controller, DSMC is the discrete-time sliding mode controller, MSVPWM is a modified space vector pulse width modulation inverter, v_{ref} is the reference load voltage, i^*_{cmd} is the desired inverter current command, i_{cmd} is the true inverter current command, v^*_{pwm} is the PWM voltage command, and v_{pwm} is the true modulated inverter output voltage. A current limiter is added between the two loops for over-current protection. The detailed development of each module will be described next.

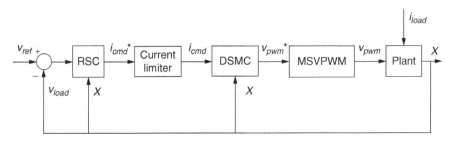

FIGURE 5.5 The proposed control system block diagram.

5.6 DESIGN OF THE DISCRETE-TIME SLIDING MODE CURRENT CONTROLLER

Discrete-time sliding mode control (DSMC) is the implementation of the sliding mode control theory in discrete time. Unlike its continuous-time counterpart, which is known for having an discontinuous control law caused by the sign function, the discrete-time sliding mode is achieved with a continuous control law. Direct implant of continuous-time sliding mode control law into discrete-time causes chattering problem due to the sampling effects. However, the discrete-time sliding mode control law does not have the problem and achieves one-step tracking for any reference input with a bandwidth lower than half of the sampling frequency given unlimited control force.

If the control force is limited, the sliding mode manifold can be reached in finite number of steps. Since the control law of the discrete-time sliding mode control is continuous from a discrete-time point of view, it can be implemented in practice with pulse width modulation with a fixed switching frequency. Continuous-time sliding mode control on an inverter typically results in varying switching frequency that may introduce more harmonics. With known plant parameters and enough control force, the discrete-time control law is equivalent to deadbeat control in formula. The most attractive property of the discrete sliding mode control is its overshoot free fast response.

For the control plant given in Eq. (5.10), the discrete-time sliding mode control law can be derived as follows. Given a sampling period of Ts and assuming zero order hold, the plant for inverter current control can be discretized as

$$X(k+1) = A_d X(k) + B_d u(k) + E_d d(k),$$
$$y(k) = C_i X(k) \tag{5.11}$$

where

$$A_d = e^{AT*}, \quad B_d = \int_0^{T_s} e^{A(T_s-T)} B \, d\tau, \quad E_d = \int_0^{T_s} e^{A(T_s-\tau)} E \, d\tau, \quad C_i = [0 \quad 1].$$

To let the current output of next step $y(k+1)$ track the reference input $i_{cmd}(k)$, a discrete-time sliding mode manifold can be chosen in the form of

$$s(k) = C_i X(k) - i_{cmd}(k),$$

that is the tracking error, such that when the discrete-time sliding mode exists, the output $y(k)$ tends to the reference $i_{cmd}(k)$. Discrete-time sliding mode can be reached if the control input $u(k)$ is designed to be the solution of

$$s(k+1) = C_i A_d X(k) + C B_d u(k) + C E_d d(k) - i_{cmd}(k) = 0, \tag{5.12}$$

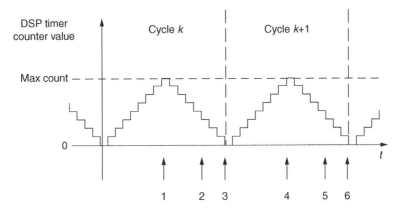

FIGURE 5.6 DSP operation causes half-sampling cycle time delay: 1, ADC of cycle k, 2, calculation done of cycle k, 3, PWM updated of cycle k, 4, ADC of cycle $k + 1$; 5, calculation done of cycle $k + 1$; 6, PWM updated of cycle $k + 1$.

The control law satisfies Eq. (5.12) is called *equivalent control* and is given by

$$u_{eq}(k) = (C_i B_d)^{-1}[i_{cmd}(k) - C_i A_d X(k) - C_i E_d d(k)], \qquad (5.13)$$

and $v^*_{pwm}(k) = u(k)$. The actual control voltage applied to the L–C filter stage is the inverter output PWM voltage $v_{pmw}(k)$, which is limited by the available top and bottom half DC bus voltages. The control force limitation caused by the DC bus voltage will be discussed later.

In practical implementation of the proposed discrete-time control strategy using a microprocessor, there exists a half sampling cycle time delay as illustrated in Fig. 5.6. Figure 5.6 shows that the discrete-time control operation of a digital signal processor (DSP) is synchronized by the system timer counting in a continuous up and down mode. When the timer count reaches its maximum (e.g., time instants 1 and 4 in Fig. 5.6), which is preset determining the sampling period, the analog-to-digital conversion (ADC) of the measured signals is triggered and the control algorithm of this cycle is launched. Based on the complexity of the control algorithm and the DSP clock frequency, the control algorithm is supposed to complete some time before the timer counts back to zero (e.g., time instants 2 and 5 in Fig. 5.6). The time duration between points 1 and 2 or 4 and 5 is the time consumed by the control calculation.

However, the updated three-phase PWM values resulting from the control calculation do not take effect until the timer counts to zero (e.g., time instants 3 and 6 in Fig. 5.6). Therefore, there is a half-cycle time delay between the time instant when the feedback variables are measured and the one when the updated PWM control command takes effect. In order to compensate for the time delay and improve the control performance, a first-order half-step predictor is introduced as follows:

$$x\left(kT_s + \frac{1}{2}T_s\right) = x(k) + \frac{x(k) - x(k-1)}{T_s} \cdot \frac{T_s}{2} = 1.5x(k) - 0.5x(k-1), \quad (5.14)$$

where x can be replaced by any discrete-time variable in the system. Therefore, the control law can be rewritten in

$$u_{eq}(k) = (C_i B_d)^{-1}\{i_{cmd}(k) - C_i A_d[1.5X(k) - 0.5X(k-1)]$$
$$-C_i E_d[1.5d(k) - 0.5d(k-1)]\}, \quad (5.15)$$

If a new vector $XDSMC$ is defined as

$$XDSMC = [i_{cmd}(k)\, v_{load}(k)\, v_{load}(k-1)\, i_{inv}(k)\, i_{inv}(k-1)\, i_{load}(k)\, i_{load}(k-1)]^{\mathrm{T}}$$

the DSMC control gain can be represented by

$$k_{DSMC} = (C_i B_d)^{-1}[1 \quad 1.5A_{d,21} \quad -0.5A_{d,21} \quad 1.5A_{d,22} \quad -0.5A_{d,22}$$
$$1.5E_{d,21} \quad -0.5E_{d,21}], \quad (5.16)$$

where $Ad;21$ denotes the second row and first column element in matrix Ad, and the control law becomes

$$u_{eq}(k) = k_{DSMC}X_{DSMC}. \quad (5.17)$$

5.7 DESIGN OF THE ROBUST SERVOMECHANISM VOLTAGE CONTROLLER

The goal of designing a realistic multivariable controller to solve the robust servomechanism problem is to achieve closed-loop stability and asymptotic regulation, as well as other desirable properties: fast response, robustness, and so on. The solution, referred to as the robustness servomechanism controller, combines both the internal model principle and the optimal control. The internal model principle says that a regulator synthesis is structurally stable only if the controller utilizes feedback of the regulated variable and incorporates in the feedback path a suitably reduplicated model of the dynamic structure of the exogenous signals which the regulator is required to process. Using the internal modeling principle for a linear time-invariant (LTI) plant, asymptotic tracking of controlled variables toward the corresponding references in the presence of disturbances (zero steady-state tracking error) can be achieved if the models that generate these references and disturbances are included in the stable closed-loop systems. A simple form of optimal control technique—linear quadratic controller—is used to obtain the feedback gain satisfying a certainly defined optimization criterion. By minimizing this criterion, the eigenvalues of the state-space model will be automatically placed and the feedback gains will be uniquely selected. The optimization criterion chosen is a functional of quadratic forms in the space of the state vector and the input of the system.

A discrete-time robust servomechanism controller (RSC) is used for outer-loop regulation. In this control system, the reference input is the desired DG unit output voltage, which is a 60-Hz sinusoidal signal, the disturbance is the load current, which may contain harmonic frequency components besides the fundamental. In practice, all high-frequency harmonics have already been suppressed by the L–C filter and the major components that may affect the system control performance are low harmonics, typically third, fifth, and seventh order harmonics, and so on. The higher the order of the harmonic, the less effect it can have on the performance. Therefore, the dynamics of the tracking or rejecting signals are known, which are governed by poles at $\pm j\omega 1$, $\pm j\omega 3$, $\pm j\omega 5$, and $\pm j\omega 7$, and so on, where $\omega 1 = 2\pi \times 60$ rad/s, $\omega 3 = 3\omega 1$, $\omega 5 = 5\omega 1$, $\omega 7 = 7\omega 1$, and so on.

The RSC is a model-based controller. Given a plant, reference signals, and disturbances, the existence of the control solution is conditional. The solution of RSC exists if the following four conditions are satisfied:

1. The control plant is stabilizable and detectable.
2. The dimension of control is greater or equal to that of the outputs.
3. The transmission zeros of the plant exclude the poles of the reference input and disturbance signals.
4. The outputs of the system is measurable.

If the above conditions are satisfied, the RSC can be designed analytically. Since the DSMC serves as the controller for in the inner loop, its dynamics has to be included together with the original plant to form the control plant for the RSC. Given the plant as shown in Eq. (5.10) and a sampling period of Ts, assuming zero-order hold and half sampling period input delay, the discretized plant is

$$X(k + 1) = A_d X(k) + B_{d0}u(k) + B_{d1}u(k - 1) + E_d d(k),$$
$$y(k) = C_v X(k) \tag{5.18}$$

where

$$A_d = e^{AT_s}, \quad B_{d0} = \int_{\frac{T_s}{2}}^{T_s} e^{A(T_s-\tau)} B\, d\tau, \quad B_{d1} = \int_0^{\frac{T_s}{2}} e^{A(T_s-\tau)} B\, d\tau,$$

$$E_d = \int_0^{T_s} e^{A(T_s-\tau)} E\, d\tau,$$

and

$$C_v = [1 \quad 0].$$

Due to the existence of $u(k - 1)$, it is reasonable to convert the plant model in Eq. (5.18) into standard state-space format as

$$\begin{aligned} X_p(k + 1) &= A_p X_p(k) + B_p u(k) + E_p d(k), \\ y(k) &= C_p X_p(k), \end{aligned} \tag{5.19}$$

where

$$X_p(k) = [X(k)^T \quad u(k - 1)]^T = [v_{load}(k) \quad i_{inv}(k) \quad u(k - 1)]^T,$$

$$A_p = \begin{bmatrix} A_d & B_{d1} \\ 0_{1 \times 2} & 0 \end{bmatrix},$$

$$B_p = \begin{bmatrix} B_{d0} \\ 1 \end{bmatrix},$$

$$E_p = \begin{bmatrix} E_d \\ 0 \end{bmatrix},$$

and

$$Cp = [C_v \, 0].$$

After the dynamics of the DSMC shown in Eq. (5.13) is included, the overall plant for the RSC is

$$\begin{aligned} X_p(k + 1) &= A_p^* X_p(k) + B_p^* u(k) + E_p d(k), \\ y(k) &= C_p X_p(k), \end{aligned} \tag{5.20}$$

where

$$A_p^* = A_p - B_p (C_i B_d)^{-1} C_i A_d C_1,$$

and

$$B_p^* = B_p (C_i B_d)^{-1},$$

where

$$C_1 = \begin{bmatrix} 1 & 0 & 0 \\ 0 & 1 & 0 \end{bmatrix}.$$

The plant for RSC described by Eq. (5.20) can be observed and proved controllable and observable using the controllability and observability criteria in linear systems theory with practically chosen circuit parameters R, L, and C. Since the tracking and rejecting dynamics have poles only at $\pm j\omega 1$, $\pm j\omega 3$, $\pm j\omega 5$, $\pm j\omega 7$, and so on, the transmission zeros of the plant can easily avoid coinciding them. In practice, this is not a concern. The input control signal to the plant (i.e., the output of the RSC) is the inverter current command. The output of the plant, for the load voltage control purpose, is the load voltage itself. Therefore, the input and output have the same dimension. In practical systems, the load voltages, the inverter currents, and the load currents are all measured. Hence, it can be declared that all of the four existing conditions of the RSC are satisfied.

The RSC design includes two parts: a servocompensator and a stabilizing compensator. The servocompensator can be designed as follows. If the tracking/disturbance poles to be considered are $\pm j\omega 1$, $\pm j\omega 3$, $\pm j\omega 5$, and $\pm j\omega 7$, the servocompensator is

$$\dot{\eta} = A_s \eta + B_s e, \tag{5.21}$$

where

$$\eta = \begin{bmatrix} \eta_1 \\ \eta_3 \\ \eta_5 \\ \eta_7 \end{bmatrix},$$

$$A_s = \begin{bmatrix} A_{s1} & & & \\ & A_{s3} & & \\ & & A_{s5} & \\ & & & A_{s7} \end{bmatrix},$$

and

$$B_s = \begin{bmatrix} B_{s1} \\ B_{s3} \\ B_{s5} \\ B_{s7} \end{bmatrix},$$

and voltage regulation error $e = v_{ref} - v_{load}$, where

$$\eta_i = \begin{bmatrix} \eta_{i1} \\ \eta_{i2} \end{bmatrix}, \quad A_{si} = \begin{bmatrix} 0 & 1 \\ -\omega_i^2 & 0 \end{bmatrix}, \quad \text{and} \quad B_{si} = \begin{bmatrix} 0 \\ 1 \end{bmatrix} \tag{5.22}$$

for $i = 1, 3, 5, 7$.

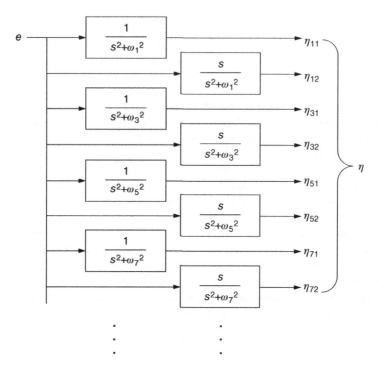

FIGURE 5.7 Block diagram of the servo compensator.

The transfer function representation of the servo compensator is illustrated in Figure 5.7, which shows that the servo compensator consists of a series of resonant filters with their resonant frequencies equal to the specified tracking or rejecting frequencies. The frequency domain characteristics of the four included resonant filter are shown in Figs. 5.8–5.11. The resonance characteristic only allows the signals with specified frequencies to pass, i.e., since the input to the servo compensator is the voltage regulation error, the voltage regulation tends to be sensitive to only those regulation error signals with specified frequencies. This idea is the same as the *resonant regulator* concept reviewed in previous section.

Equation (5.22) is the most straightforward state-space implementation of a resonant filter with a specified frequency. However, in practice, the values of the elements in matrix A_{si} may significantly differ from each other, which may cause a poor condition number of the matrix and numerical instability in microprocessor implementation, especially for fixed-point microprocessors. To overcome this problem, an orthogonal transformation can be applied to the matrices A_{si} and B_{si} to balance up the elements in A_{si} while keeping the transfer function unchanged. This transformation can be described as $A_{si}, bal = T^{-1} A_{si} T$ and $B_{si}, bal = T^{-1} B_{si}$, where the orthogonal matrix T can be obtained using MATLABR function ssbal().

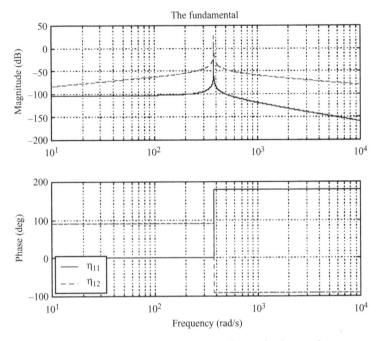

FIGURE 5.8   Bode plot of servocompensator for the fundamental component.

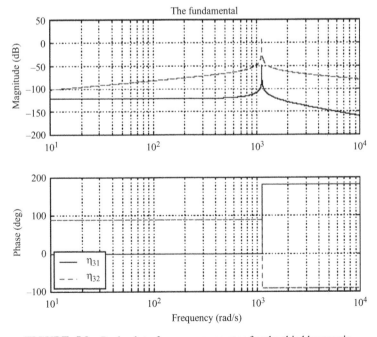

FIGURE 5.9   Bode plot of servocompensator for the third harmonic.

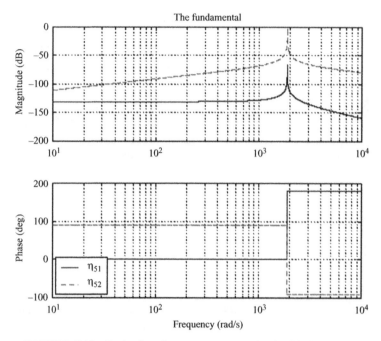

FIGURE 5.10   Bode plot of servocompensator for the fifth harmonic.

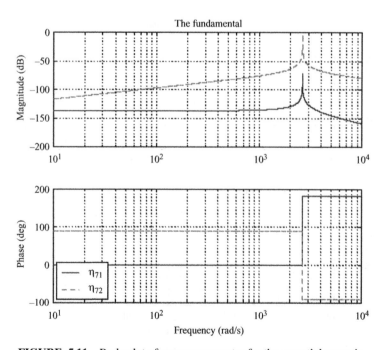

FIGURE 5.11   Bode plot of servocompensator for the seventh harmonic.

The servocompensator needs to be discretized in practice. Given the same sampling period Ts as above, the discrete-time servocompensator is

$$\eta(k+1) = A_{sd}\eta(k) + B_{sd}e(k), \tag{5.23}$$

where

$$A_{sd} = e^{A_s T_s} \quad \text{and} \quad B_{sd} = \int_0^{T_s} e^{A_s(T_s - \tau)} B_s \, d\tau.$$

With the existence of the servocompensator, the stabilizing compensator can be generated. An augmented system combining both the plant [Eq. (5.20)] and the servocompensator [Eq. (5.23)] can be written as

$$\hat{X}(k+1) = \hat{A}\hat{X}(k) + \hat{B}u_1(k) + \hat{E}_1 d(k) + \hat{E}_2 y_{ref}(k), \tag{5.24}$$

where the control input is $u1(k) = i_{cmd}(k)$, the load disturbance is $d(k) = i_{load}(k)$, the reference input is $y_{ref}(k) = v_{ref}(k)$, the state vector is $\hat{X} = [X_p^T \quad \eta^T]^T$, and the coefficient matrices are

$$\hat{A} = \begin{bmatrix} A_p^* & 0 \\ -B_{sd}C_p & A_{sd} \end{bmatrix},$$

$$\hat{B} = \begin{bmatrix} B_p^* \\ 0 \end{bmatrix},$$

$$\hat{E}_1 = \begin{bmatrix} E_p \\ 0 \end{bmatrix},$$

and

$$\hat{E}_2 = \begin{bmatrix} 0 \\ B_{sd} \end{bmatrix}.$$

The task of the stabilizing compensator is to stabilize the augmented system in Eq. (5.24). To achieve this goal, the discrete-time LQ optimal control technique can be used, which guarantees stability of the system while yielding optimized performance by minimizing a discrete-time linear quadratic performance index

$$J_\varepsilon = \sum_{k=1}^{\infty} \hat{X}(k)^T Q \hat{X}(k) + \varepsilon u_1(k)^T u_1(k), \tag{5.25}$$

where Q is a symmetrical positive-definite matrix to be chosen and $\varepsilon > 0$ is a small number to reduce the weight of the control force in the optimization. Obtaining the state feedback gain K minimizing J_ε requires solving the algebraic Riccati equation

$$\hat{A}^T P + P\hat{A} + Q - \frac{1}{\varepsilon} P\hat{B}\hat{B}^T P = 0 \tag{5.26}$$

for the unique positive semidefinite solution P, so that

$$K = -\frac{1}{\varepsilon}\hat{B}^T P. \tag{5.27}$$

Then the control input can be obtained as

$$u_1(k) = K^T \hat{X} = [K_0 \quad K_1] \begin{bmatrix} X_p \\ \eta \end{bmatrix}. \tag{5.28}$$

The block diagram of the RSC is shown in Fig. 5.12. In practice, the discrete-time algebraic Riccati equation can be solved using the MATLABR function dlqr(). Since the system is linear time-invariant, the feedback gain K is a constant value calculated in advance and does not change in operation. Therefore, online updating of K is unnecessary. In practice, the selection of Q has significant impacts on the control performance. In the proposed technique, three different gains, w_p, w_1, and w_h, are used as weights for the plant states (i.e., X_p), fundamental servocompensator states (i.e., $\eta 11$ and $\eta 12$), and harmonic servo compensator states (i.e., $\eta 31$, $\eta 32$, $\eta 51$, $\eta 52$, $\eta 71$, and $\eta 72$), respectively. Therefore the weight matrix can be obtained as

$$Q = \begin{bmatrix} Q_1 & 0 \\ 0 & Q_2 \end{bmatrix}, \tag{5.29}$$

where $Q_1 = w_p I_3$, and

$$Q_2 = \begin{bmatrix} w_1 I_2 & 0 \\ 0 & w_h I_6 \end{bmatrix},$$

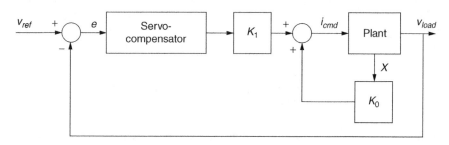

FIGURE 5.12   Block diagram of the RSC.

where I_n denotes an n-dimensional identity matrix. To make good use of the servocompensator, w_p should be significantly less than w_1 and w_h. To emphasize good fundamental tracking, w_1 can be set significantly greater than w_h.

5.8 LIMIT THE CURRENT COMMAND

The current command signal $i_{cmd}^*(k) = u_1(k)$ generated by the RSC needs to be limited to perform overload protection. In the discussed DG inverter system topology, current limit I_{max} is given as the maximum peak current allowed for a phase. However, the voltage and current controls are conducted in the stationary $\alpha\beta0$ reference frame. To solve this problem, a new current limiting algorithm has been developed as follows:

$$
i_{cmd,i} = \begin{cases} i_{cmd,i}^*, & \text{for } \sqrt{i_{cmd,\alpha}^{*2} + i_{cmd,\beta}^{*2}} + |i_{cmd,0}^*| \leq I_{max}, \\ \dfrac{I_{max}}{\sqrt{i_{cmd,\alpha}^{*2} + i_{cmd,\beta}^{*2}} + |i_{cmd,0}^*|} i_{cmd,i}^*, & \text{for } \sqrt{i_{cmd,\alpha}^{*2} + i_{cmd,\beta}^{*2}} + |i_{cmd,0}^*| > I_{max}, \end{cases}
$$

$$(5.30)$$

where the subscript $i \in \{\alpha, \beta, 0\}$, $i*_{cmd,\alpha}$, $i*_{cmd,\beta}$, and $i*_{cmd,0}$ are the current commands generated by the RSCs for each axis.

To prevent servocompensator states (which are related to the current command) from growing while the current command is saturated, the following strategy can be applied. Rewrite the servocompensator equation as

$$\eta(k+1) = A_{sd}\eta(k) + B_{sd}e_1(k), \tag{5.31}$$

where

$$
e_1(k) = \begin{cases} e(k), & \text{for } \sqrt{i_{cmd,\alpha}^{*2} + i_{cmd,\beta}^{*2}} + |i_{cmd,0}^{*2}| \leq I_{max}, \\ 0, & \text{for } \sqrt{i_{cmd,\alpha}^{*2} + i_{cmd,\beta}^{*2}} + |i_{cmd,0}^{*2}| > I_{max}. \end{cases}
$$

When the current command is limited, the voltage control loop is open and the system is running under current-controlled mode.

5.9 A MODIFIED SPACE VECTOR PWM

The conventional space vector PWM technique pulse width modulation (PWM) is a process that varies the time duration of a sequence of pulses according to a reference signal to get a reference-signal-modulated pulse sequence. In power electronics, PWM techniques are typical tools to control high-power voltage or current waveforms using low-power signals. The so-called space vector PWM (SVPWM)

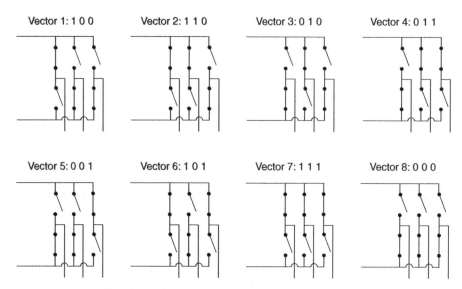

FIGURE 5.13 Switching patterns of space vector PWM.

is a known and widely used three-phase pulse width modulation technique, which is originally developed for three-phase three-wire topologies where the inverter does not provide a neutral point. The basic operating principle of the conventional SVPWM for three-phase voltage source inverters is as follows. The SVPWM is a specially designed switching sequence for the power switches in a three-phase inverter using base space vectors to generate three-phase 120± apart sinusoidal line-to-line output voltages. This technique treats the sinusoidal reference voltage as a constant amplitude vector rotating at a certain frequency. There are eight base space vectors in total, corresponding to eight possible switching patterns of the inverter. The switching patterns of the inverter and associated base space vectors are shown in Fig. 5.13. Each of the three elements reflects the on/off state of one of the three legs in the inverter. If the inverter has a Y connected load as shown in Fig. 5.14, the output line-to-neutral and line-to-line voltages are shown in Table 5.1 assuming normalized DC bus voltage. Among the eight base vectors, vectors 1 to 6 generate nonzero output voltages and the three-phase voltage waveforms generated by repeated sequences of Vector 1 to 6 are illustrated in Fig. 5.15, where the line-to-neutral voltages are known as the six-step waveform. Vectors 7 and 8 generate zero output voltage in the topology of Fig. 5.14 by connecting all three outputs to the upper or bottom rail of the DC bus.

The fundamental components of the waveforms shown in Fig. 5.15 are three-phase 120°-apart sine waves. However, there are significant amount of harmonics due to the step waveforms caused by the base vectors which do not provide intermediate values in between the steps. Weighting of these discrete base vector values in time duration allows intermediate values to be obtained. Weighting the duration of the base vectors

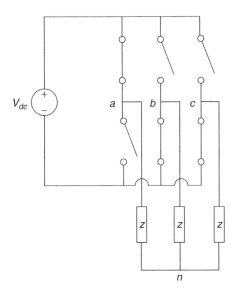

FIGURE 5.14   Three-phase three-wire inverter topology with Y connected load.

according to a reference vector is the main idea of SVPWM. The base vectors can be depicted geometrically as shown in Fig. 5.16, where the consecutive nonzero vectors on a plane form six 60° sectors representing a full sinusoidal period matching the result in Fig. 5.15. A sinusoidal reference signal can be represented by a vector on the same plane rotating from one sector to another with a constant speed. In each sector, the reference vector can be achieved by time-averaging the two boundary base vectors and two zero vectors with proper weights, as illustrated in Fig. 5.17 showing the case of Sector I.

TABLE 5.1   Output Voltage Patterns of Base Vectors Under Normalized DC Bus Voltage

	a	b	c	v_{an}	v_{bn}	v_{cn}	v_{ab}	v_{bc}	v_{ca}
Vector 1	1	0	0	$\frac{2}{3}$	$-\frac{1}{3}$	$-\frac{1}{3}$	1	0	-1
Vector 2	1	1	0	$\frac{1}{3}$	$\frac{1}{3}$	$-\frac{2}{3}$	0	1	-1
Vector 3	0	1	0	$-\frac{1}{3}$	$\frac{2}{3}$	$-\frac{1}{3}$	-1	1	0
Vector 4	0	1	1	$-\frac{2}{3}$	$\frac{1}{3}$	$\frac{1}{3}$	-1	0	1
Vector 5	0	0	1	$-\frac{1}{3}$	$-\frac{1}{3}$	$\frac{2}{3}$	0	-1	1
Vector 6	1	0	1	$\frac{1}{3}$	$-\frac{2}{3}$	$\frac{1}{3}$	1	-1	0
Vector 7	1	1	1	0	0	0	0	0	0
Vector 8	0	0	0	0	0	0	0	0	0

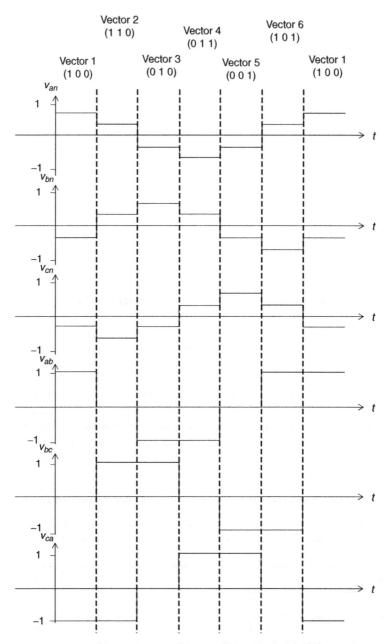

FIGURE 5.15 Output voltage waveforms of space vector PWM base vectors.

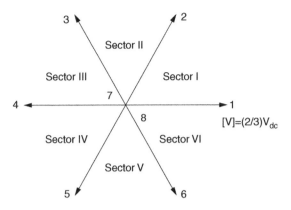

FIGURE 5.16   Base vectors of space vector PWM in two-dimensional space.

Similar to the well-known sine wave PWM, SVPWM is also a carrier-based fixed switching frequency modulation technique, where each power switch is turned on and off a fixed number of times in a period of time. In any one PWM period, a switch is turned on and off once. The PWM switching frequency should be significantly greater than the reference frequency to generate enough number of pulses and reduce harmonics. In Fig. 5.17, T_z is half of the PWM period, T_1 is the duration of Vector 1, and T_2 is the duration of Vector 2 in a half-PWM period. The remaining time $T_z - T_1 - T_2$ is the duration for the zero vectors. The values of T_1 and T_2 are calculated based on the values of V_{ref} and μ_{ref}. The advantages of conventional SVPWM for a three-phase

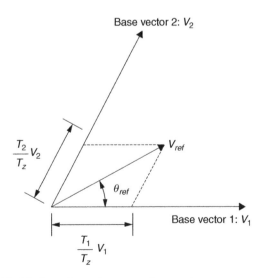

FIGURE 5.17   Modulation of a reference voltage using space vector PWM in two-dimensional space.

TABLE 5.2   Output Voltage Patterns of Base Vectors in the Four-Wire Split DC Bus Topology

	a	b	c	v_{an}	v_{bn}	v_{cn}	v_{ab}	v_{bc}	v_{ca}
Vector 1	1	0	0	$\frac{1}{2}$	$-\frac{1}{2}$	$-\frac{1}{2}$	1	0	-1
Vector 2	1	1	0	$\frac{1}{2}$	$\frac{1}{2}$	$-\frac{1}{2}$	0	1	-1
Vector 3	0	1	0	$-\frac{1}{2}$	$\frac{1}{2}$	$-\frac{1}{2}$	-1	1	0
Vector 4	0	1	1	$-\frac{1}{2}$	$\frac{1}{2}$	$\frac{1}{2}$	-1	0	1
Vector 5	0	0	1	$-\frac{1}{2}$	$-\frac{1}{2}$	$\frac{1}{2}$	0	-1	1
Vector 6	1	0	1	$\frac{1}{2}$	$-\frac{1}{2}$	$\frac{1}{2}$	1	-1	0
Vector 7	1	1	1	$\frac{1}{2}$	$\frac{1}{2}$	$\frac{1}{2}$	0	0	0
Vector 8	0	0	0	$-\frac{1}{2}$	$-\frac{1}{2}$	$-\frac{1}{2}$	0	0	0

three-wire system include less harmonic distortion in output voltages and/or currents applied to the phases and more efficient use of the supply voltage in comparison with direct sinusoidal modulation technique, that is, higher DC bus utilization.

5.9.1   A MODIFIED SPACE VECTOR PWM TECHNIQUE

In three-phase three-wire systems, the three phases are not independent, that is, there exist only two independent degrees of freedom. The conventional SVPWM only provides two-dimensional control, and that is why the base vectors are on the same plane. However, in a three-phase four-wire system with split DC bus topology discussed, the neutral of the load is grounded, which yields three independent phases and calls for three-dimensional control. In this topology, Vectors 7 and 8 are no longer zero vectors in terms of line-to-neutral voltages. The output voltages of the split DC bus inverter are shown in Table 5.2, and the corresponding waveforms are shown in Fig. 5.18.

From Table 5.2, it can be observed that Vectors 7 and 8 provide nonzero line-to-neutral voltages which comprise the third dimension, the 0-axis quantity, not existing in the three-wire topology. The 0-axis voltages or currents in the three-phase circuit caused by Vector 7 or 8 are called common mode voltages or currents because they are in phase to each other.

A modified space vector PWM technique (MSVPWM) has been developed to take advantage of the Vectors 7 and 8 to achieve three-dimensional control in an $\alpha\beta0$ system. In conventional SVPWM, the durations of Vectors 7 (000) and 8 (111) are equal in a half-PWM period, as shown in Fig. 5.19, and there is no 0-axis control capability. However, in MSVPWM, unequal duration of Vectors 7 and 8 causes uneven 0-axis average voltage in a PWM period and allows 0-axis current to flow, which enables the inverter to perform control in all α, β, and 0 axes. A Sector I illustration is shown in Fig. 5.20.

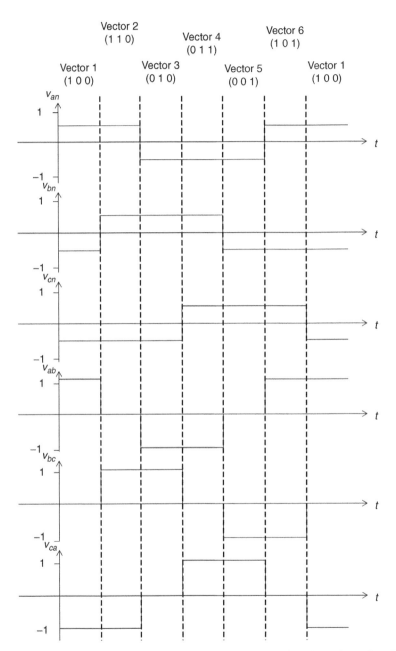

FIGURE 5.18 Output voltage waveforms of the base vectors in a four-wire split DC bus topology.

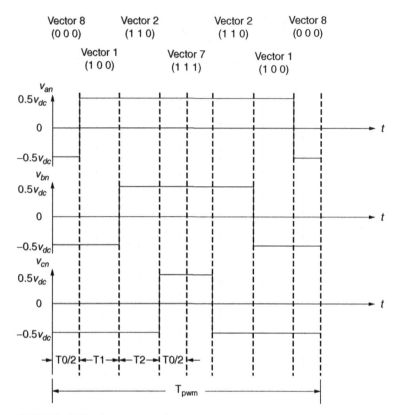

FIGURE 5.19 Sector I on–off sequence of conventional space vector PWM.

Calculation of the vector durations in MSVPWM is shown below, using Section I as an example. Given three dimensional reference voltage inputs, v_α, v_β, v_0, and the PWM period T_{pwm}, the α–β plane reference voltage is

$$V_{ref} = \sqrt{v_\alpha^2 + v_\beta^2},$$ (5.32)

its phase angle is

$$\theta_{ref} = \arctan \frac{v\beta}{v\alpha},$$ (5.33)

and the modulation index is

$$a = \frac{V_{ref}}{\frac{1}{2}V_{DC}}.$$ (5.34)

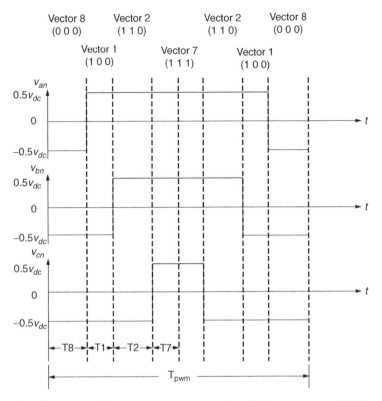

FIGURE 5.20 Sector I on–oα sequence of modified space vector PWM.

As mentioned above, $T_z = (1/2)T_{pwm}$ is the half-PWM period. The Vector 1 duration is

$$T_1 = \frac{\sin\left(\dfrac{\pi}{3} - \theta_{ref}\right)}{\sin\dfrac{\pi}{3}} T_z a \tag{5.35}$$

and the Vector 2 duration is

$$T_2 = \frac{\sin\theta_{ref}}{\sin\dfrac{\pi}{3}} T_z a. \tag{5.36}$$

The sum of the Vector 7 and 8 durations is $T_0 = T_z - T_1 - T_2$. After defining

$$\Delta T = \frac{v_0}{\dfrac{1}{2}V_{dc}} T_z,$$

the durations of Vectors 7 and 8 are

$$T_7 \begin{cases} \dfrac{1}{2}(T_0 + \Delta T) & \text{for } |\Delta T| \le T_0, \\ T_0 & \text{for } |\Delta T| > T_0, \\ 0 & \text{for } |\Delta T| < -T_0 \end{cases} \tag{5.37}$$

and

$$T_8 \begin{cases} \dfrac{1}{2}(T_0 + \Delta T) & \text{for } |\Delta T| \le T_0, \\ 0 & \text{for } |\Delta T| > T_0, \\ T_0 & \text{for } |\Delta T| < -T_0. \end{cases} \tag{5.38}$$

The one PWM period time average voltage waveforms comparing the conventional SVPWM and the new MSVPWM have been shown in Fig. 5.21, where the dashed curve is the line-to-neutral voltage generated by conventional SVPWM while the solid curve is that by MSVPWM with a 0-axis compensation injection shown by the dotted line. In a feedback control system, this compensation signal $v0$ can be automatically generated by the controller in the feedback loop.

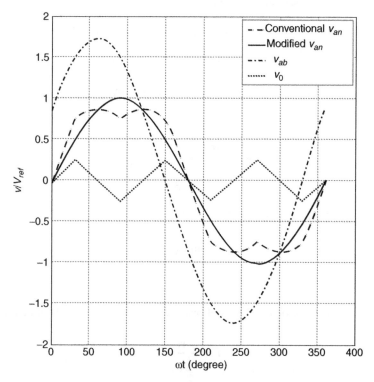

FIGURE 5.21   Output waveforms of the space vector PWM techniques.

The benefit of this MSVPWM is that it provides three-dimensional control for three-phase four-wire split DC bus topology inherently in the stationary $\alpha\beta 0$ reference frame, which performs better than direct ABC reference frame control when the DC bus voltage is limited.

Further analysis results verifying this statement will be presented later. The trade-off of this technique is that it has less DC bus voltage utilization in that the maximum reference voltage in the conventional SVPWM is $V_{ref.max} = \dfrac{\sqrt{3}}{3}V_{DC}$ and the MSVPWM is $V_{ref;max} = (1/2)V_{DC}$.

5.10 PERFORMANCES AND ANALYSIS—FREQUENCY DOMAIN ANALYSIS

In this section, both frequency domain analysis and time domain simulation and experiments will be presented.

A 5-KVA DG unit with three-phase four-wire split DC bus topology as shown in Fig. 5.1 is analyzed. The rectifier is assumed to supply DC voltages regulated at ± 270 V. The rated RMS line-to-neutral three-phase AC output voltage is 120 V. The discrete-time control frequency, as well as the PWM switching frequency, $fs = 1/Ts$ is set to be 5.4 kHz.

A three-phase L–C filter is selected with $Lf = 10.2$ mH and $Cf = 55$ μF. This filter has a natural frequency of 1335.1 rad/sec and damping ratio of 0.0367, where a line/coil resistance of 1 is assumed. The bode plot of this filter is shown in Fig. 5.22. Greater values of L and C yield better filter performance, but large L leads to high weight and volume and large C leads to high capacitor current at no load. Therefore the selection of L and C is usually a trade-off and the values can be selected to yield a natural frequency or cutoff frequency lower than 10th harmonic frequency, which filters out high-frequency components, leaves lower frequency components for the control to handle, and keeps the weight and volume within a reasonable range. In the split bus topology, the filter input voltage (i.e., v_{pwm}) could never be zero, which tends to cause high switching frequency current. In order to suppress this current, a higher L is necessary especially when the switching frequency is not high. When the current-control loop is closed while the voltage loop remains open, the frequency response of the current-controlled system is shown in Fig. 5.23 and the corresponding discrete-time pole-zero map on Z-plane is shown in Fig. 5.24. The responsiveness of the closed-loop control can be measured by bandwidth, which is the frequency where the gain falls to -3 dB. In Fig. 5.23, it can be observed that the closed-loop system has a bandwidth greater than 10,000 rad/sec, which explains the fast response of the current control. Ideally, without the input time delay caused by the microprocessor, the bandwidth of DSMC reaches half of the sampling frequency.

In Fig. 5.24, all closed-loop poles are in the unit circle, so that the current control loop is stable.

The closed-loop voltage control has a frequency response shown in Fig. 5.25. It can be observed that with the existence of the RSC, all tracking/rejecting frequencies have a unity gain which is desired.

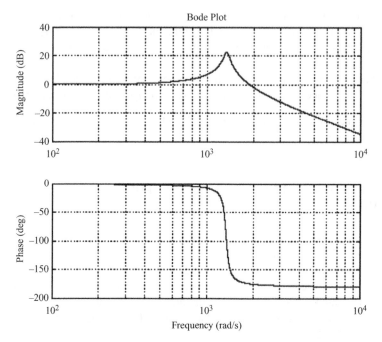

FIGURE 5.22   Bode plot of the *L–C* filter.

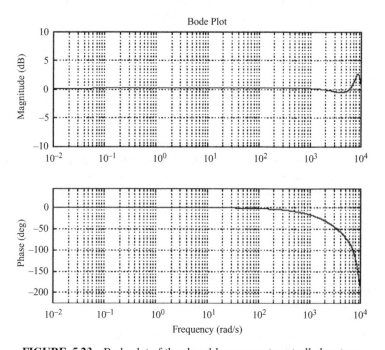

FIGURE 5.23   Bode plot of the closed-loop current-controlled system.

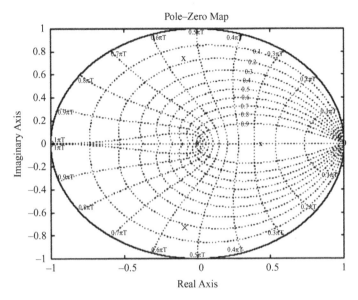

FIGURE 5.24 Poles and zeros of the closed-loop current-controlled system.

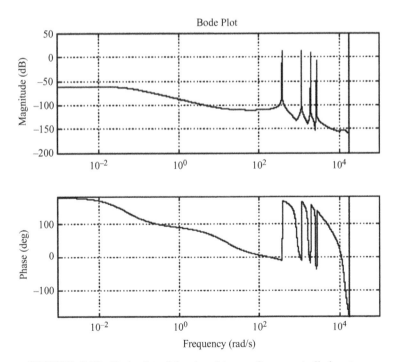

FIGURE 5.25 Bode plot of the closed-loop voltage-controlled system.

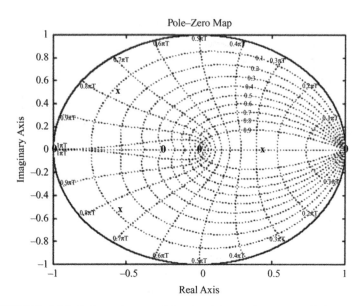

FIGURE 5.26 Poles and zeros of the closed-loop voltage-controlled system.

The discrete-time pole-zero map on Z-plane is shown in Fig. 5.26 where all closed-loop poles are in the unit circle, which indicates that the voltage control loop is stable.

5.10.1 Time Domain Simulations

Time domain simulations of the closed system have been performed under various load scenarios, including both steady state and transient. A computer simulation software MATLAB® with Simulink SimPowerSystems Blockset is used. Steady-state RMS output voltages and THDs under different types of load are presented in Table 5.3. The load types include 5-kVA balanced resistive load, unbalanced resistive loads applied on phase A only and on both phases A and B, 5-kVA inductive load with 0.8 power factor, no load, and nonlinear load with crest factor 2:1, where the crest factor is equal to the peak amplitude of a waveform divided by the RMS value. It can be observed from the table that the proposed control technique produces nearly zero steady-state error and THD lower than 1% for all tested load types.

The output waveforms of load voltages, load currents, inverter currents, RMS load voltages, and ground current in the above scenarios are shown in Figs. 5.27–5.31. In the balanced load scenarios, the 0-axis ground current, including the capacitor current, does not have low frequency components besides switching frequency. However, in the unbalanced load cases, the ground current has a fundamental component and in the nonlinear load case, it has a third harmonic dominant AC component. The transient response of the control has been demonstrated in Figs. 5.32 and 5.33, where

TABLE 5.3   Simulation Results of Steady-State Performances of the Proposed Control Technique

Load Type	RMS Load Voltage (V)			THD
	$V_{load.A}$	$V_{load.B}$	$V_{load.B}$	
Resistive load	120.0	120.0	120.0	0.80%
Unbalanced A	120.0	120.0	120.0	0.47%
Unbalanced A and B	120.0	120.0	120.0	0.52%
Inductive load 4	120.0	120.0	120.0	0.50%
No load	120.0	120.0	120.0	0.70%
Nonlinear load	120.3	120.3	120.5	0.98%

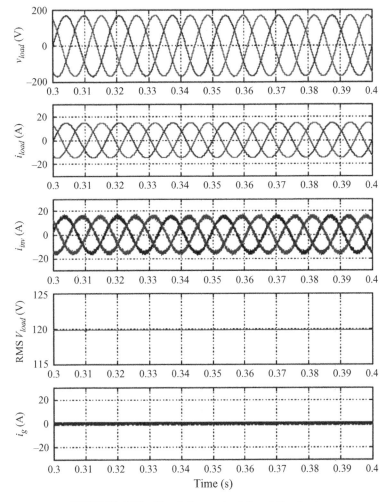

FIGURE 5.27   Simulation under full resistive load.

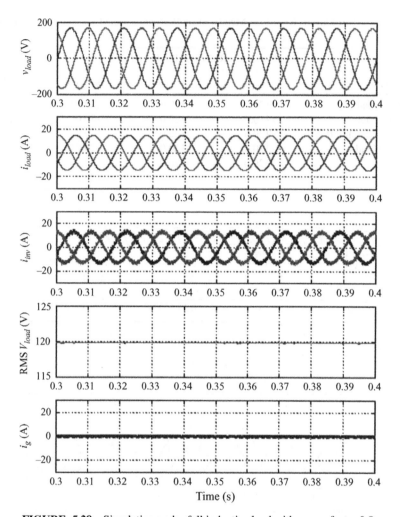

FIGURE 5.28   Simulation under full inductive load with power factor 0.8.

a full resistive load is instantaneously applied to or removed from the out-put terminal. The figures show that the voltage waveforms are only slightly affected by the load transients and restored to steady state in a very short period of time; that is, the waveform dents last for only 2 msec, and the RMS values show about 2-V deviations on each transient, which last for about 20 msec.

5.11   EXPERIMENTAL RESULTS—THE EXPERIMENTAL SETUP

Experimental tests of the proposed control technique have been performed on a 1-kVA prototype system with ±200 V DC bus, and a greater inductance $Lf = 60.2$ mH

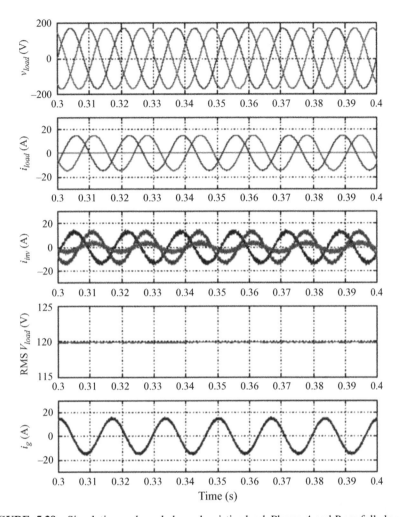

FIGURE 5.29 Simulation under unbalanced resistive load: Phases A and B are fully loaded.

is used to provide strong suppression of the 0-axis switching frequency current which affects control performance significantly in practical systems. The filter capacitance $Cf = 55$ μF remains the same as above. The power converter uses SEMIKRON® SKM 50GB 123D IGBT modules with SKHI 22 gate drives. A Texas Instruments® TMS320LF2407A digital signal processor (DSP) with a Spectrum Digital® evaluation module (EVM) has been used as the digital controller. The block diagram of the setup is shown in Fig. 5.34. In Fig. 5.34, the voltage divider is used to sense high voltages by lowering it down to within a ±12-V range which is operable by operational amplifiers. LEM® LA55- P current output Hall effect current transducers are used to measure currents with a 75–0.1% burden resistor for each channel.

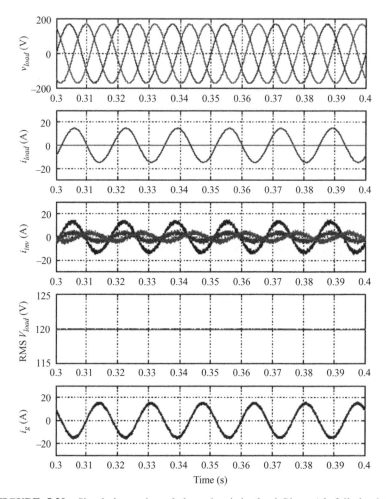

FIGURE 5.30 Simulation under unbalanced resistive load: Phase *A* is fully loaded.

The signal conditioning stage adjusts the input measurement signals into 0 to 3.3-V range with 1.65 V DC offset matching the requirement of the analog-to-digital conversion of the DSP. Texas Instruments® TL074BCN four-channel operational amplifiers are used in the signal conditioning. The optical isolation interfaces the DSP pulse-width modulation output pins with the IGBT gate drives. The isolation prevents the DSP from possible damages caused by fault in the power stage. Dual-channel Hewlett Packard® HCPL-2232 optocouplers are used.

The DC bus overvoltage protection circuit protects the DC electrolytic capacitors and the IGBT modules from possible overvoltage fault caused by the front end. Once overvoltage is detected, the front-end power supply will be shut down and a DC bus discharging circuit will be switched on to lower the DC bus voltage. The test waveforms have been presented in Figs. 5.35–5.44. Although the experimental

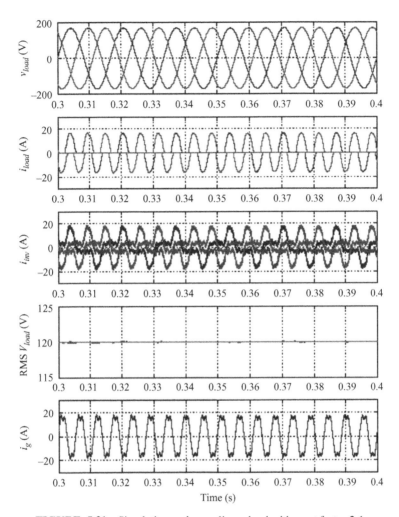

FIGURE 5.31 Simulation under nonlinear load with crest factor 2:1.

results are less perfect compared to the simulation ones, they are reasonably good to demonstrate the effectiveness of the proposed control technique. The experimental results can be improved by using faster PWM switching frequency, which reduces switching frequency current and allows higher control gain leading to better control performance.

5.11.1 Stationary $\alpha\beta0$ Reference Frame Versus *ABC* Reference Frame

One simulation research has been performed comparing the performances of conducting the controls in stationary $\alpha\beta0$ reference frame versus *ABC* reference frame. It is natural to use the MSVPWM if the $\alpha\beta0$ reference frame is used and to use

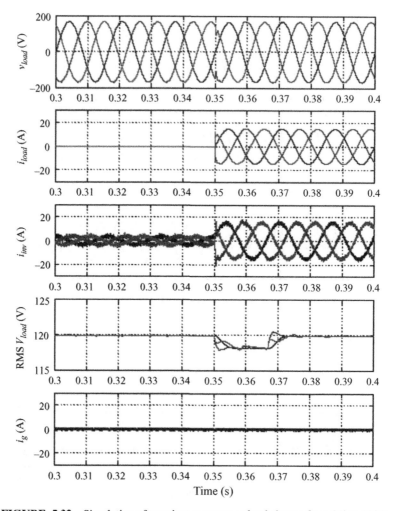

FIGURE 5.32   Simulation of transient response at load change from 0% to 100%.

conventional sine wave PWM if the *ABC* reference is used. As stated before, mutual-transformable relationship can be constructed between the conventional SVPWM and USPWM which is a sine wave PWM technique in a three-phase three-wire system. It is imaginable that there exists a similar relationship between the MSVPWM and sine wave PWM while the detailed proof deserves further investigation. If this is true and the harmonic issues in the switching frequency and vicinity are ignored, conditional equivalence between the two mod-ulation strategies in linear modulation area can be conceivable in terms of control performance point of view. In other words, in linear modulation area, the control performances of the two modulation strategies are comparable. If this statement is true again, it can be demonstrated that the statement

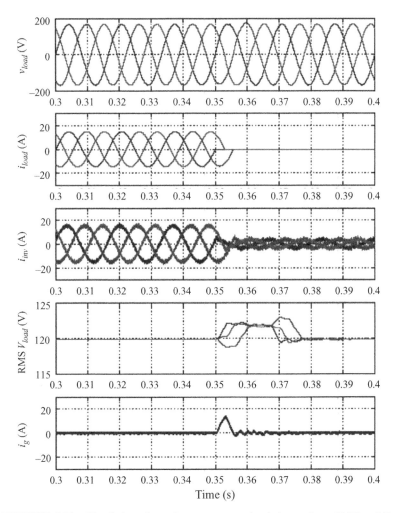

FIGURE 5.33   Simulation of transient response at load change from 100% to 0%.

becomes untrue when the modulation index approaches 1 due to the limited DC bus voltage or high modulation references.

When the modulation index a approaches 1, as shown by the algorithm, $T_1 + T_2$ approaches T_0 and the space for 0-axis modulation tends to zero. This strategy implies a higher priority for the modulation of $\alpha-\beta$ axes than that of the 0-axis when the DC bus voltage is limited. In other words, when the 0-axis modulation is saturated at T_0, the modulation of $\alpha-\beta$ references can still be linear. This is not the case in sine wave PWM, where the total modulation reference is saturated as the modulation index approaches 1 given the condition that overmodulation is not allowed due to its negative impact on the modulation.

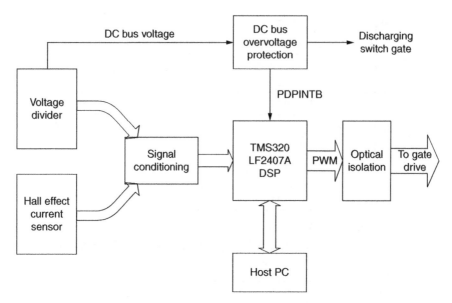

FIGURE 5.34   Block diagram of the experimental setup.

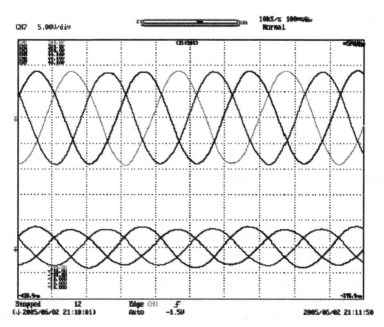

FIGURE 5.35   Experimental result of load voltages and currents under resistive load. **Upper**: 100 V/div. **Lower**: 1 A/div.

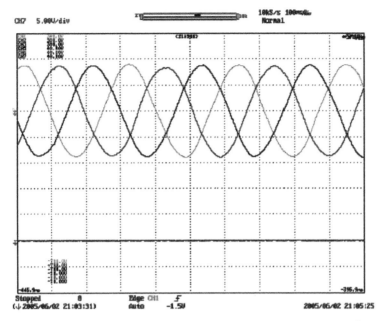

FIGURE 5.36 Experimental result of load voltages and currents under no load. **Upper**: 100 V/div. **Lower**: 1 A/div.

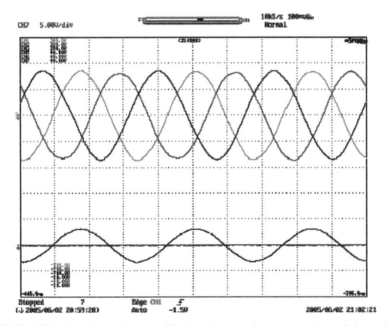

FIGURE 5.37 Experimental result of load voltages and currents under unbalanced load. Phase A is loaded. **Upper**: 100 V/div. **Lower**: 1 A/div.

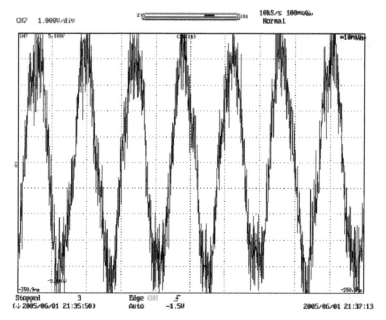

FIGURE 5.38 Experimental result of ground current under unbalanced load. Phase A is loaded, 1 A/div.

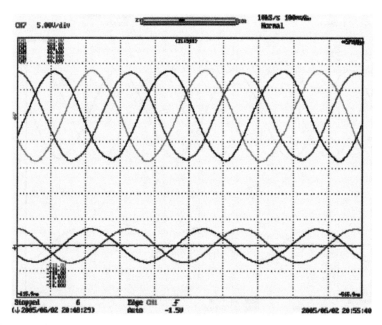

FIGURE 5.39 Experimental result of load voltages and currents under unbalanced load, Phases A and B are loaded. **Upper**: 100 V/div. **Lower**: 1 A/div.

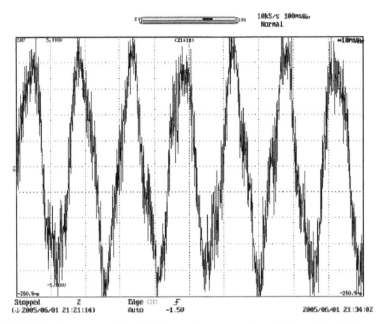

FIGURE 5.40 Experimental result of ground current under unbalanced load. Phases A and B are loaded, 1 A/div.

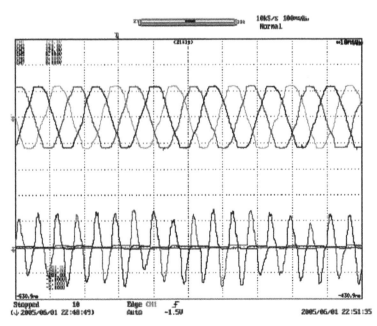

FIGURE 5.41 Experimental result of load voltages and currents under nonlinear load. **Upper**: 40 V/div. **Lower**: 1 A/div.

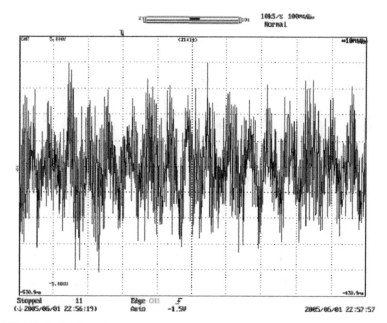

FIGURE 5.42   Experimental result of ground current under nonlinear load, 1 A/div.

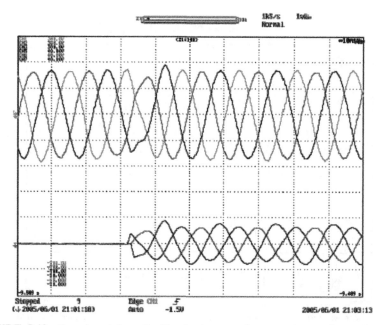

FIGURE 5.43   Experimental result of load voltages and currents in stepping up load transient. **Upper**: 100 V/div. **Lower**: 1 A/div.

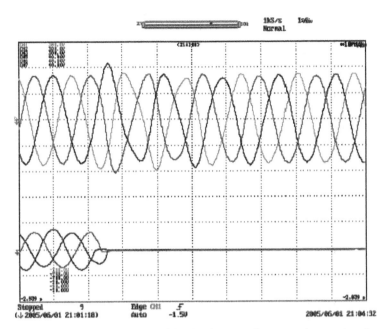

FIGURE 5.44 Experimental result of load voltages and currents in stepping down load transient. **Upper**: 100 V/div. **Lower**: 1 A/div.

Simulation results have demonstrated that the above difference about the reference frames does matter in the control performance. In the simulations using two different reference frames, the same load, the same L–C filter, the same DC bus voltage ±257.5 V, and the same controllers, are used for the comparison purpose. Figure 5.45 shows the THD performances of the two. The one using $\alpha\beta0$ reference frame and MSVPWM yields steady low THD consistently while the one using ABC reference frame and sine-wave PWM ends up with rising THD and losing stability.

The above comparison indicates that the control in the stationary $\alpha\beta0$ reference frame with MSVPWM to allows use of lower DC bus voltage than the one using ABC reference frame and sine wave PWM. This property is desirable in control of DG units with the three-phase four-wire split DC bus topology because the relatively low DC bus voltage utilization of this topology tends to ask for better usage of the DC bus voltage to avoid using very high DC bus voltage for cost concerns.

5.12 THE ROBUST STABILITY: BASIC IDEAS ABOUT UNCERTAINTY, ROBUST STABILITY, AND μ- ANALYSIS

The proposed DG control technique, including the RSC and DSMC controllers, is a model-based control. However, the mathematical model can never precisely describe the true physical process. The mismatch between the model and the true process

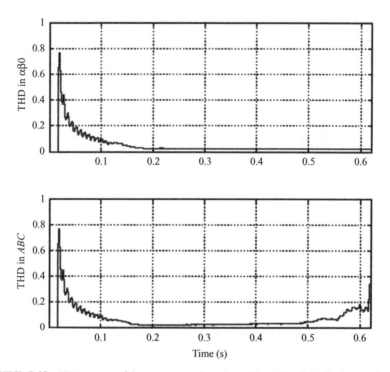

FIGURE 5.45 THD curves of the same control conducted in $\alpha\beta0$ and ABC reference frames.

is caused by unmodeled dynamics, which is the so-called uncertainty. Uncertainty usually contains two sources: parametric uncertainty and external disturbances. The uncertainty is unknown in practice, but it is reasonable to assume that its boundary is known, which enables mathematical analysis. There are two typical ways to model uncertainty: to model it as external inputs or as parametric perturbation to a nominal model. In this section, the latter approach will be used. There are also two ways to analyze uncertainty: to analyze it as unstructured uncertainty or structured uncertainty. Unstructured uncertainty refers to the case where all sources of uncertainty are lumped into a single perturbation. With this strategy, a single boundary of all perturbations is used which yields very conservative analysis result in terms of robust stability. Structured uncertainty refers to the case where the individual sources of uncertainty are described separately. This approach increases the complexity of the analysis but gives a less conservative result. Structured uncertainty-based analysis will be used in this section.

As stated before, a feedback control system is said to achieve robust stability if it remains stable for all considered perturbations caused by uncertainty in the plant. Structured singular value μ can be used to analyze the stability robustness of a multi-input–multi-output (MIMO) linear system under structured uncertainty. In order to do so, an uncertainty model needs to be constructed as shown in Fig. 5.46, where M represents a state-space representation of transfer matrix of a known stable MIMO linear system with a dimension $n \times n$ and Δ is a structured uncertainty matrix typically with block diagonal format including perturbation values of all uncertainty sources. Define set

$$\mathbf{\Delta} = \{\mathrm{diag}[\delta_1 I_{r1}, \ldots, \delta_S I_{rS}, \Delta_1, \ldots, \Delta_F] : \delta_i \in \mathbb{C}, \Delta_j \in \mathbb{C}^{m_j \times m_j}\}, \qquad (5.39)$$

where $\delta_i I_i$ denotes a repeated scalar block and Δ_j denotes a full block, integers S and F represent the number of repeated scalar blocks and the number of full blocks, respectively, and $\mathbb{C}$ denotes the complex set. Then $\Delta \in \mathbf{\Delta}$ and

$$\sum_{i=1}^{S} r_i + \sum_{j=1}^{F} m_j = n.$$

For $M \in \mathbb{C}^{n \times n}$, the structured singular value is defined as

$$\mu_{\mathbf{\Delta}}(M) \overset{\Delta}{=} \frac{1}{\min\{\bar{\sigma}(\Delta) : \Delta \in \mathbf{\Delta}, \det(I - M\Delta) = 0\}}, \qquad (5.40)$$

where $\bar{\sigma}(\Delta)$ is the largest singular value of Δ, unless no $\Delta \in \mathbf{\Delta}$ makes $I - M\Delta$ singular, in which case $\mu_{\Delta}(M) \overset{\Delta}{=} 0$.

The robust stability of system in Fig. 5.46 is achieved for all Δ with $||\Delta||_\infty < 1$ if and only if

$$\sup \mu \Delta(M) \leq 1. \qquad (5.41)$$

The value of μ is difficult to obtain in a real system, while the upper bound can be calculated using MATLAB$^\circledR$ function mu()in the robust control toolbox.

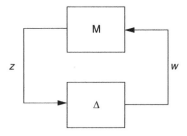

FIGURE 5.46   Generalized uncertainty model for robust stability analysis.

5.12.1   Linear Fractional Transformation and Uncertain Open-Loop Model

In order to perform μ-analysis on the robust stability of the system, an uncertain model with the structure shown in Fig. 5.46 needs to be constructed. Uncertainty is modeled as perturbations to a nominal model. In the DG unit discussed in this research, the perturbations include load disturbances. Therefore, the load model should be included in the plant model also, as shown in Fig. 5.47, where all parameters are in per-unit values, and admittance model is used for modeling an inductive load; that is, $Y_G = 1/R_{load}$, is the complement of the parallel load resistance and $Y_B = 1/L_{load}$ is the complement of the parallel load inductance. Nonlinear load is not considered in this analysis.

The parametric uncertainty of a component is usually bounded by its manufacturing tolerance. The manufacturing tolerances of the discussed circuit components are as follows:

- Filter inductor tolerance: $\pm 15\%$,
- Filter inductor loss tolerance $\pm 50\%$, and
- Filter capacitor tolerance $\pm 6\%$.

The nominal load admittance is 1. Assuming a power factor of 0.8 lagging, the conductance is 0.8 and the susceptance is 0.6. The specification of the DG unit requires a stable operation range without performance degradation under load from 0 to 200%. This is equivalent to conductance variation from 0 to 1.6 and susceptance variation from 0 to 1.2. Therefore $Y_{G,nom} = 0.8$ and $Y_{B,nom} = \omega 1 B_{nom} = 2\pi \times 60 \times 0.6 = 226.19$ with the same tolerance of $\pm 100\%$.

Therefore, the circuit parameters can be redefined by adding perturbations on top of their nominal values defined as shown below:

$$L = L_{nom}(1 + L_{tol}\delta L), \ |\delta L| < 1, \tag{5.42}$$

$$R = R_{nom}(1 + R_{tol}\delta R), \ |\delta R| < 1, \tag{5.43}$$

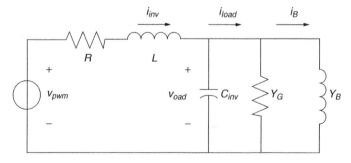

FIGURE 5.47   One-dimensional equivalent circuit for robust stability analysis.

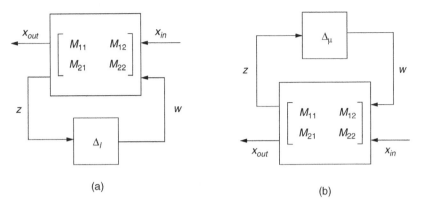

(a)   (b)

FIGURE 5.48 (A) A lower LFT (B) An upper LFT.

$$C = C_{nom}(1 + C_{tol}\delta C), \; |\delta C| < 1, \tag{5.44}$$

$$Y_G = Y_{G,nom}(1 + Y_{G,tol}\delta G), \; |\delta G| < 1, \tag{5.45}$$

and

$$YB = Y_{B,nom}(1 + Y_{B,tol}\delta B), \; |\delta B| < 1, \tag{5.46}$$

where L_{nom}, R_{nom}, and C_{nom} equal to the per-unit circuit parameters L, R, and C, respectively, and $L_{tol} = 0.15$, $R_{tol} = 0.5$, $C_{tol} = 0.06$, $Y_{G,tol} = 1$, and $Y_{B,tol} = 1$.

The dynamic model of the equivalent circuit in Fig. 5.48 with redefined parameters where the perturbations are included can be represented by

$$\dot{v}_{load} = -\frac{Y_G}{C}v_{load} + \frac{1}{C}i_{inv} - \frac{1}{C}i_B, \tag{5.47}$$

$$\dot{i}_{inv} = -\frac{1}{L}v_{load} - \frac{R}{L}i_{inv} + \frac{1}{L}v_{pwm}, \tag{5.48}$$

and

$$\dot{i}_B = Y_B v_{load}. \tag{5.49}$$

With the parameter perturbations defined as in equations 5.47 through 5.49, the linear fractional transformations (LFT) has to be performed to construct an uncertain open-loop model. The purpose for LFT is to pull the perturbations out and make them separate from the nominal model while keeping their interconnections with their own sources.

This can be explained as follows. Let M be a complex matrix partitioned as shown in Figure 5.48 and let $\Delta_\ell \in \mathbb{C}^{q_2 \times p_2}$ and $\Delta_u \in \mathbb{C}^{q_1 \times p_1}$ be two other complex matrices.

A lower LFT with respect to Δ_ℓ can be defined as

$$\mathcal{F}_\ell(M, \Delta_\ell) \triangleq M_{11} + M_{12}\Delta_\ell(I - M_{22}\Delta_\ell)^{-1}M_{21}, \tag{5.50}$$

provided that the inverse $(I - M_{22}\Delta_\ell)^{-1}$ exists. An upper LFT with respect to Δu can be defined as

$$\mathcal{F}_u(M, \Delta_u) \triangleq M_{22} + M_{21}\Delta_u(I - M_{11}\Delta_u)^{-1}M_{12}, \tag{5.51}$$

provided that the inverse $(I - M_{11}\Delta_u)^{-1}$ exists. A lower and an upper LFT are illustrated in Figs. 5.48(A) and 5.48(B) respectively, where

$$\begin{bmatrix} x_{out} \\ z \end{bmatrix} = M \begin{bmatrix} x_{in} \\ w \end{bmatrix} = \begin{bmatrix} M_{11} & M_{12} \\ M_{21} & M_{22} \end{bmatrix} \begin{bmatrix} x_{in} \\ w \end{bmatrix},$$

$w = \Delta_\ell z$ for the lower LFT and

$$\begin{bmatrix} z \\ x_{out} \end{bmatrix} = M \begin{bmatrix} w \\ x_{in} \end{bmatrix} = \begin{bmatrix} M_{11} & M_{12} \\ M_{21} & M_{22} \end{bmatrix} \begin{bmatrix} w \\ x_{in} \end{bmatrix},$$

$w = \Delta u z$ for the upper LFT. In the construction of the uncertain open-loop model, only the lower LFT is used. In Eqs. (5.47)–(5.49), perturbed parameters L and C appear in denominators.

A lower LFT of L is

$$\frac{1}{L} = \frac{1}{L_{nom}(1 + L_{tol}\delta_L)} = \frac{1}{L_{nom}} - \frac{L_{tol}}{L_{nom}}\delta_L(1 + L_{tol}\delta_L)^{-1} = \mathcal{F}_\ell(M_L, \delta_L),$$

where

$$M_L = \begin{bmatrix} \dfrac{1}{L_{nom}} & -L_{tol} \\ \dfrac{1}{L_{nom}} & -L_{tol} \end{bmatrix}.$$

Similarly, $\mathcal{F}_\ell(M_C, \delta_C)$ can be obtained with

$$M_C = \begin{bmatrix} \dfrac{1}{C_{nom}} & -C_{tol} \\ \dfrac{1}{C_{nom}} & -C_{tol} \end{bmatrix}.$$

As for the perturbed parameters R, Y_G, and Y_B which are in numerators, their lower LFTs can be obtained as

$$M_R = \begin{bmatrix} R_{nom} & \Delta R \\ 1 & 0 \end{bmatrix},$$

$$M_{Y_G} = \begin{bmatrix} R_{G,nom} & \Delta Y_G \\ 1 & 0 \end{bmatrix},$$

and

$$M_{Y_B} = \begin{bmatrix} Y_{B,nom} & \Delta Y_B \\ 1 & 0 \end{bmatrix},$$

where $\Delta R \triangleq R_{nom} \cdot R_{tol}$, $\Delta Y_G \triangleq Y_{G,nom} \cdot Y_{G,tol}$, and $\Delta Y_B \triangleq R_{B,nom} \cdot Y_{B,tol}$. These LFTs can be illustrated in Figs. 5.50 and 5.51. After interconnecting all modules in Figs. 5.49 and 5.50 based on the relation in Eqs. (5.47)–(5.49), the system structured perturbation can be illustrated in Fig. 5.51.

The dynamics in Fig. 2.52 can be represented by the following augmented equation:

$$Y_{aug} = P_{aug} U_{aug}, \tag{5.52}$$

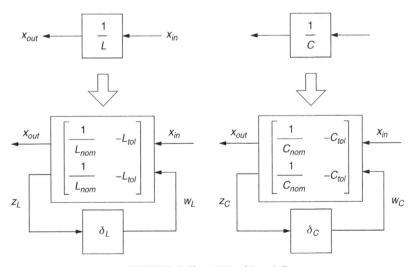

FIGURE 5.49   LFTs of L and C.

where

$$Y_{aug} = \begin{bmatrix} \dot{v}_{load} \\ \dot{i}_{inv} \\ \dot{i}_B \\ \dot{v}_{load} \\ i_{inv} \\ i_{load} \\ z_C \\ z_L \\ z_R \\ z_{Y_B} \\ z_{Y_G} \end{bmatrix},$$

$$P_{aug} = \begin{bmatrix} -\dfrac{Y_{G,nom}}{C_{nom}} & \dfrac{1}{C_{nom}} & -\dfrac{1}{C_{nom}} & 0 & -C_{tol} & 0 & 0 & 0 & -\dfrac{\Delta Y_G}{C_{nom}} \\[2.2ex] -\dfrac{1}{L_{nom}} & -\dfrac{R_{nom}}{L_{nom}} & 0 & \dfrac{1}{L_{nom}} & 0 & -L_{tol} & -\dfrac{\Delta R}{L_{nom}} & 0 & 0 \\[2.2ex] Y_{B,nom} & 0 & 0 & 0 & 0 & 0 & 0 & \Delta Y_B & 0 \\[1.5ex] 1 & 0 & 0 & 0 & 0 & 0 & 0 & 0 & 0 \\[1.5ex] 0 & 1 & 0 & 0 & 0 & 0 & 0 & 0 & 0 \\[1.5ex] Y_{G,nom} & 0 & 1 & 0 & 0 & 0 & 0 & 0 & \Delta Y_G \\[2.2ex] -\dfrac{Y_{G,nom}}{C_{nom}} & \dfrac{1}{C_{nom}} & -\dfrac{1}{C_{nom}} & 0 & -C_{tol} & 0 & 0 & 0 & -\dfrac{\Delta Y_G}{C_{nom}} \\[2.2ex] -\dfrac{1}{L_{nom}} & -\dfrac{R_{nom}}{L_{nom}} & 0 & \dfrac{1}{L_{nom}} & 0 & -L_{tol} & -\dfrac{\Delta R}{L_{nom}} & 0 & 0 \\[2.2ex] 0 & 1 & 0 & 0 & 0 & 0 & 0 & 0 & 0 \\[1.5ex] 1 & 0 & 0 & 0 & 0 & 0 & 0 & 0 & 0 \\[1.5ex] 1 & 0 & 0 & 0 & 0 & 0 & 0 & 0 & 0 \end{bmatrix},$$

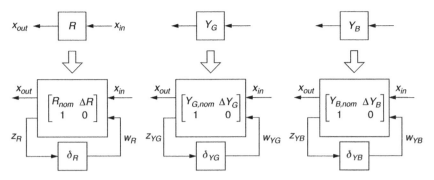

FIGURE 5.50   LFTs of R, Y_G, and Y_B.

and

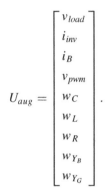

$$U_{aug} = \begin{bmatrix} v_{load} \\ i_{inv} \\ i_B \\ v_{pwm} \\ w_C \\ w_L \\ w_R \\ w_{Y_B} \\ w_{Y_G} \end{bmatrix}.$$

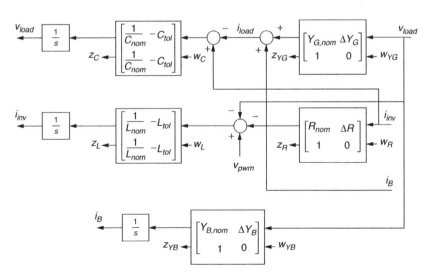

FIGURE 5.51   Uncertain open-loop model with structured perturbations.

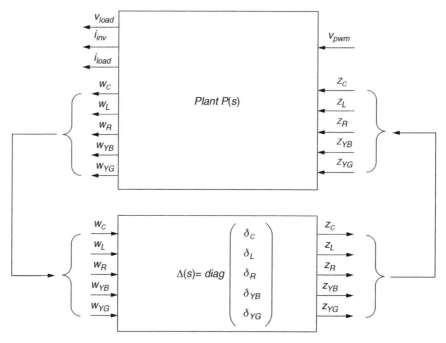

FIGURE 5.52   Uncertain open-loop model showing the perturbation block matrix.

The LFTs allow the perturbation block matrix Δ to be pulled out from the nominal model while the perturbation structures remain through the interconnections:

$$W = \{w_C, w_L, w_R, w_{YB}, w_{YG}\} \quad \text{and} \quad Z = \{z_C, z_L, z_R, z_{YB}, z_{YG}\},$$

as shown in Figs. 5.52 and 5.53, where v_{pwm} and $Y = \{v_{load}, i_{inv}, i_{load}\}$ are the external input and output. The model represented in Figs. 5.52 and 5.53 are the uncertain open-loop model.

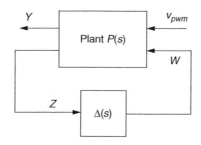

FIGURE 5.53   Uncertain open-loop model: A higher-level block diagram.

5.12.2   Uncertain Closed-Loop Model

In order to study the impacts of the perturbations on the closed-loop system stability, the controllers have to be taken into consideration. With the external input and output ports, v_{pwm} and v_{load}, i_{inv}, i_{load}, given in Fig. 5.52, the controllers can be interconnected with the uncertain open-loop model to form an uncertain closed-loop model.

The dynamics of of the controllers, the DSMC and RSC can be constructed using the designed control parameters. MATLABR with its Control System Toolbox and μ-Analysis and Synthesis Toolbox has been utilized to conduct the interconnection and conversions between continuous time and discrete time. The following steps are necessary to obtain the closed-loop plant model:

1. Apply zero-order hold to the continuous plant $P(s)$, to include the effect of the sample and hold process of the digital sampling process.

2. Transform the discretized plant back to the continuous system by applying an inverse Tustin transformation, which has the property of preserving the frequency response of the discrete-time systems.

3. Obtain state-space representations of the RSC and DSMC with inputs and outputs definitions as shown in Fig. 5.54. The combined controller state-space system, containing both the RSC and DSMC, can be calculated using MATLAB$^{\circledR}$ command sysic().

4. The combined controller system is then transformed into a complex plane in continuous domain using the inverse Tustin transformation.

5. Finally, the closed-loop plant model M is obtained by invoking the command sysic() again to complete the interconnections between the combined controller and the uncertain open-loop model to obtain an uncertain closed-loop model.

The uncertain closed-loop model is shown in Figs. 5.55 and 5.56, where the latter is a high-level diagram resembling the standard format for studying robust stability of systems with uncertainty as shown in Fig. 5.46.

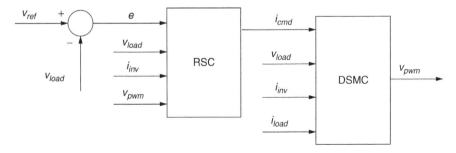

FIGURE 5.54   Block diagram of the controllers showing the input and output ports.

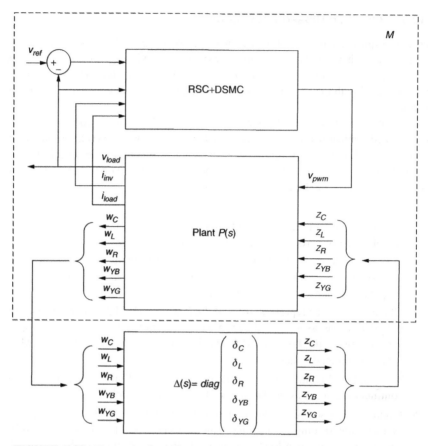

FIGURE 5.55   Uncertain closed-loop model showing the perturbation block matrix.

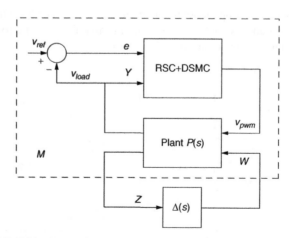

FIGURE 5.56   Uncertain closed-loop model: A higher-level block diagram.

5.12.3 Robust Stability and Gain Tuning

With the existence of uncertain closed-loop model, it is possible to conduct robust stability analysis under structured perturbations using the structured singular value $\mu\Delta(M)$ and study the impact of control gains to the robust stability.

The control parameters obtained for the DSMC are fixed, and there is no need to tune them while those of the RSC can be tuned. The control gain K is obtained by solving the algebraic Riccati equation in Eq. (5.26). The solution of the Riccati equation is dependent on the weight matrix Q, which is a function of weights w_p, w_1, and w_h. Therefore, K can be tuned by adjusting w_p, w_1, and w_h. Different K values yield different control performances. An example is given in Fig. 5.57, which is a simulation comparison of the RMS transient responses of the output voltage v_{load} to the step-up load (0–100%) and step-down load (100%–0) disturbances under three different w_p values—that is, 0.5, 0.05, and 0.005 respectively, where the values of w_1 and w_h are fixed at 5×10^7 and 5×10^5, respectively.

In the simulation, the weight for the control commands is set to 10^{-5}. Since the linear quadratic control is practiced here, it is straightforward that the lower the weight on one part of the performance index as shown in Eq. (5.25), the higher the gain for the other part of the performance index. Therefore, lower w_p leads to higher gain on the servo states, which yields better tracking/rejecting performance at the specified frequencies. This statement can be verified by observing the curves in Fig. 5.57. The figure shows that negative variations of the RMS output voltage are caused under the step-up load transient while positive variations are caused under the step-down case. In both cases, the gain $w_p = 0.005$ leads to the least variations among the three, which is the result of its highest servo gain. However, the above study is conducted without

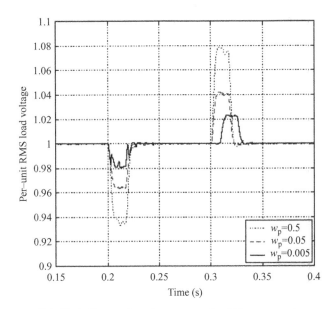

FIGURE 5.57 RMS performance under different w_p values.

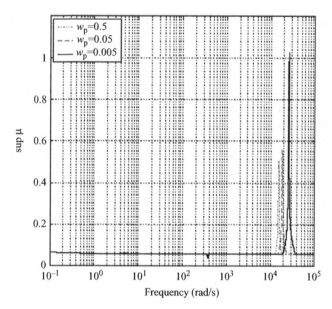

FIGURE 5.58   Upper bound of $\mu\Delta(M)$ under different w_p values.

any perturbations. If the perturbations as defined in Eqs. (5.42)–(5.46) are applied, the robust stability issue will be raised. The robust stability of the perturbed system under different gains has been studied by performing μ-analysis on the system. The robust stability can be checked by observing the upper bound of $\mu\Delta(M)$ over the frequency span of interest as shown in Fig. 5.58. In this figure, $w_p = 0.005$ makes the upper bound of μ exceed the robust stability criterion (i.e., sup $\mu < 1$), at $\omega \approx 25,$ 000 rad/sec, which means unstable. However, the other weight values $w_p = 0.05$ and $w_p = 0.5$ give sup $\mu < 1$ for the entire frequency span and the lower servo gain case; that is, $wp = 0.5$ has more stability margin.

The existing five perturbation sources do not share the contribution to the instability evenly. A study on the contributions of individual perturbations to the same system as studied above has been conducted by calculating frequency responses $\dfrac{|z_i(j\omega)|}{|\omega_i(j\omega)|}$, where $i = 1, \ldots, 5$, at $w_p = 0.5$. The resulted frequency responses have been plotted in Fig. 5.59, where traces 1–5 are the cases with perturbation only on C, L, R, Y_B, and Y_G, respectively. These plots agree with the robust stability analysis result given above that the system achieves the robust stability because all five magnitude responses have a gain less than 1, which is the criterion. However, these five cases yield different robust stability margin. The response to the filter inductance perturbation is the one closest to 1, which yields the least robust stability margin, given the tolerances defined in Section 5.12. The robust stability of the closed-loop system under parametric perturbations on the filter inductance has been investigated by finding the positions of the poles representing the dynamics of the inductance under various amount of perturbation

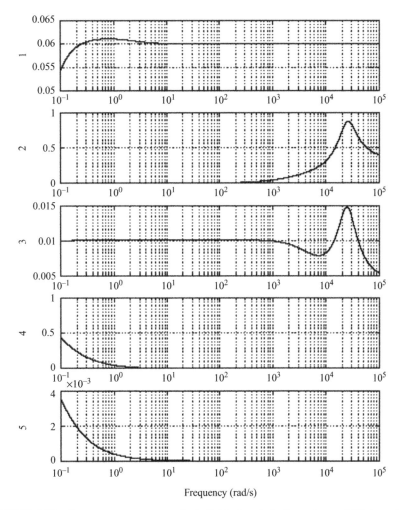

FIGURE 5.59   Frequency responses under individual perturbations at $w_p = 0.05$, magnitude only, that is, $\dfrac{|z_i(j\omega)|}{|\omega_i(j\omega)|}$, where $i = 1, \ldots, 5$: 1, perturbation on C; 2, perturbation on L; 3, perturbation on R; 4, perturbation on Y_B; and 5, perturbation on Y_G.

measured by the value of δL. The pole-zero map is presented in Fig. 5.60, where the poles under $\delta_L \in \{0, -0.5, -1.0, -1.5, -2.0, -2.5\}$ have been shown. It can be observed that the closed loop tends to be unstable when the value of δL decreases in that the poles drift to the right half of the complex plane.

In this chapter, a new control technique for a three-phase four-wire DG unit with split DC bus topology has been proposed. This technique is a combination of discrete-time sliding mode control (DSMC) and robust servomechanism control (RSC). The development of the control algorithm have been presented based on a

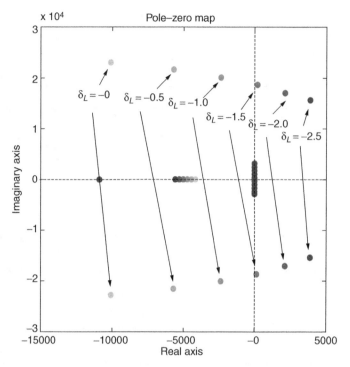

FIGURE 5.60 Closed-loop pole and zero map with $w_p = 0.05$ and various δL values, including both stable and unstable cases.

one-dimensional equivalent circuit model of the system in stationary $\alpha\beta 0$ reference frame and per-unit values.

A modified space vector PWM technique (i.e., the MSVPWM) has been proposed to perform three-dimensional control on an $\alpha\beta 0$ basis. The pulse duration algorithm yielding higher priority on the non-zero-axis dimensions (i.e., α and β) under limited DC bus voltage has been presented. Simulation comparison has shown the advantage of this approach over the conventional sine wave PWM in the *ABC* reference frame. A series of analysis and studies have been performed on the proposed control technique, including the *L–C* filter design issue, closed-current-loop and closed-voltage-loop responses, and time domain simulations and experiments under various load scenarios. All these analyses, simulations, and experiments have demonstrated the effectiveness of the proposed control solution.

The robust stability of the proposed solution in the presence of linear load disturbances and parametric uncertainty has been verified by the structured singular-value-based method. It has been shown that the weight adjustment in the optimal control performance index provides a way of tuning the transient performance of the controller while maintaining stability robustness of the system under perturbations.

CHAPTER 6

POWER FLOW CONTROL OF A SINGLE DISTRIBUTED GENERATION UNIT

In this chapter, we will use the mathematical model for a grid-connected DG unit with local load to present the control of active reactive power in island mode and a grid connected mode. As a first step, a MATLAB simulation testbed will be constructed. The control method presented will be used on the simulation testbed. Finally, we will present the experimental testbed. The experimental testbed will be used in conjunction with control technology

6.1 INTRODUCTION

A distributed generation (DG) unit with distributed energy sources, such as fuel cells, micro-turbines, and photovoltaic devices, can be connected to a utility grid, there by pumping power into the grid besides providing power to their local loads. In this chapter, the DG unit is interfaced with utility grid through a three-phase inverter. With inverter control, both real and reactive power pumped into the utility grid from the DG unit can be controlled. Real power flow P control determines whether the the utility grid is taking power from the DG or supplying the local load together with the DG and how much this power should be. Reactive power flow control allows the DG unit to be used as static var compensation unit besides an energy source. With possible existence of nonlinear local load or harmonic distorted utility voltage, the line current is still desired to be sinusoidal and harmonic free. The existing power control solutions have limitations—for example, not suitable for DG applications

Integration of Green and Renewable Energy in Electric Power Systems. By A. Keyhani, M. N. Marwali, and M. Dai
Copyright © 2010 John Wiley & Sons, Inc.

with existence of local load, slow response, coupling between P and Q control, sensitiveness of line current to harmonic distorted utility grid, and so on.

This chapter will present a DG unit real and reactive power control technique to a three-phase DG unit with three-phase three-wire topology and isolation transformer. This technique provides robust voltage regulation with harmonic elimination in island running mode. In a grid-connected mode, the control performs online power system identification estimating the Thevenin equivalent parameters of the utility grid and uses the acquired information to conduct a feedforward power control together with the traditional integral control. A harmonic power controller will be proposed to handle possible harmonic distorted utility grid voltage. The proposed technique yields fast and less coupled real and reactive power flow control with line current conditioning capability even under nonlinear local load and harmonic corrupted grid voltage in grid-connected mode.

The power control technique proposed in this chapter is based on existing voltage and current control strategy of individual DG unit control in the island model as described in Chapters 4 and 5 (which combine discrete-time sliding mode current control and robust servomechanism voltage control) but is developed for a three-wire topology. The three-wire topology has an advantage in reduced dimension of control plant and nonexistence of 0-axis current, both leading to ease of control. This is why this topology has been chosen to perform the grid-connected studies. The closed-loop P and Q control stability will be proved using the Lyapunov direct method.

6.2 THE CONTROL SYSTEM

The proposed control solution is developed for a grid-connected DG unit shown in Fig. 6.1. The DG unit consists of a DC bus powered by any DC source or AC source with a rectifier, a voltage source inverter, an L–C filter stage, and a Δ/Y_g-type isolation transformer with secondary side filtering. The DG unit has a local load, linear or nonlinear, and is connected to the utility grid through a three-phase static switch. The utility grid is modeled by Thevenin's Theorem as an equivalent three-phase AC source with an equivalent internal impedance.

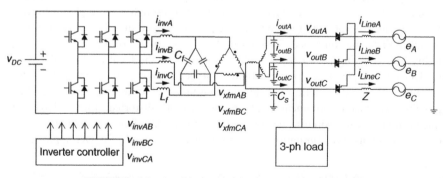

FIGURE 6.1 A grid-connected DG unit with local load.

In island mode, the inverter conducts voltage control, where the load voltage *VoutABC* should track the given reference. The voltage control goal is strong voltage regulation, low static error in RMS, fast transient response, and low total harmonic distortion (THD). If the voltage of the utility main *EABC* is measured and used as the reference, *VoutABC* will be controlled to match *EABC* in magnitude and synchronized in phase angle.

In grid-connected mode, the DG unit conducts power control, where the output active power P and reactive power Q from the DG unit to the utility grid should be regulated to desired values P_{ref} and Q_{ref}. Both P_{ref} and Q_{ref} can be positive or negative, which provides the possibility for the DG unit to help with the energy production and stability enhancement of the power system or sustain power supply to local load when it exceeds the capacity of the DG. The control goal of power regulation is stability, low static error, and fast response with low coupling between P and Q. A conventional solution is integral control, forcing the DG unit to perform like a large generator with fast dynamics. This approach is a purely feedback control not requiring any plant knowledge. However, with the measurable variables and existing computing power of digital signal processors (DSP), it is possible to acquire some useful plant information that can play an instructive role in developing power controller.

In practice, the utility grid voltage can often be harmonic distorted or the local load can draw harmonic current from the unit. These situations make it necessary to have the power controller to handle the problem and yield low harmonic line power and sinusoidal line current. The power flow control strategy will be presented in detail here in this chapter, which addresses these issues and provides solutions.

As stated above, the power flow control technique developed in this research is on the basis of a three-phase three-wire version of RSC plus DSMC voltage and current control solution similar to the technique discussed in Chapter 2. For completeness and for considering the difference between the three-wire and four-wire topology, the main attributes of the voltage and current controls will still be reviewed.

6.3 VOLTAGE AND CURRENT CONTROL

For high quality of $V_{out}ABC$ with strong regulation, low THD, and overload protection, a dual-loop voltage and current control structure is used as shown in Fig. 6.2, where the inner loop is for current control while the outer loop is for voltage control. A robust servomechanism controller (RSC) is used for voltage control, and a discrete-time sliding mode controller (DSMC) is used for current control. The three-phase quantities are transformed from an ABC reference frame into a stationary $\alpha\beta$ reference frame since there is no 0-axis current involved in control.

The DSMC is used in the current loop to limit the inverter current under overload condition because of the fast and overshoot-free response it provides. The RSC is adopted for voltage control due to its capability to perform zero steady-state tracking error under unknown load and eliminate harmonics of any specified frequencies with guaranteed system stability. The theory behind the RSC is based on the solution of robust servomechanism problem (RSP), where the internal model principle and the

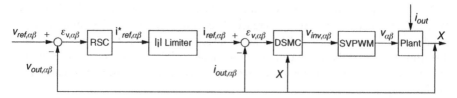

FIGURE 6.2 Control structure for island mode.

optimal control theory for linear systems are combined. The internal model principle is applied in the DG voltages control by including the fundamental frequency mode and the frequency modes of the harmonics to be eliminated into the controller. The linear quadratic LQ optimal control theory of linear systems is combined in the RSC in order to guarantee the stability of the closed-loop system and providing arbitrary good transient response based on desired control priorities.

In a DG unit shown in Fig. 6.1, most of the voltage harmonics, like the triplets (3rd, 9th, 15th, ...) and even harmonics, are either nonexisting or uncontrollable, or negligible in value. Therefore, only the fundamental and the 5th and 7th harmonics are left for the control to handle. Since the overload protection is a strongly desired feature for inverter systems, a DSMC is included in the inner loop to limit the current under overload conditions. With the existence of the DSMC in the inner loop, the RSC in the outer loop are designed taking the dynamics of the DSMC into account, so that the stability and robustness of the overall control system is guaranteed.

6.3.1 The Discrete-Time Sliding Mode Current Control

The circuit shown in Fig. 6.1 is a linear system and can be modeled in state space. The current controller controls the inverter current—that is the inductor current. The L–C filter can be represented in discrete time as

$$\{X_i(k+1) = A_i X_i(k) + B_i v_{inv}(k) + E_i i_{out}(k), \ y_i(k) = C_i X_i(k) \quad (6.1)$$

where $X_i(k) = [v_{xfm}^T(k) \ i_{inv}^T(k)]^T$, v_{inv}, and load disturbance i_{out} are all represented in $\alpha\beta$ reference frame. Given inverter current command $i_{ref}(k)$, the DSMC equivalent control law is

$$v_{inv,eq}(k) = (C_i B_i)^{-1} [i_{ref}(k) - C_i A_i X_i(k) - C_i E_i i_{out}(k)]. \quad (6.2)$$

With inverter control voltage limit v_{max} determined by the DC bus voltage and PWM technique, the actual control voltage becomes

$$v_{inv} = \begin{cases} v_{inv,eq} & \text{if } \|v_{inv,eq}\| \le v_{max}, \\ \dfrac{v_{max}}{\|v_{inv,eq}\|} v_{inv,eq} & \text{otherwise.} \end{cases} \quad (6.3)$$

6.3.2 The Robust Servomechanism Voltage Control

The voltage controller generates current command i_{ref} for the current controller. For the system shown in Fig. 6.1, considering the dynamics of the DSMC current controller, the overall state-space model is

$$X(k + 1) = AX(k) + B i_{ref}(k) + E i_{out}(k), \tag{6.4}$$

where $X(k) = [v_{xfm}^T(k)\ i_{inv}^T(k)\ v_{out}^T(k)\ i_{xfm}^T(k)]^T$, with all element vectors represented under $\alpha\beta$ reference frame, and i_{xfm} is the transformer secondary current and will be replaced by i_{out} feedback in the control due to the negligible difference.

An RSC is a combination of a stabilizing compensator and a servocompensator. The stabilizing compensator is the state feedback of Eq. (6.4) times a gain $K0$. The servocompensator is a second order filter represented by

$$\eta(k + 1) = A_s\eta(k) + B_s e_v(k), \tag{6.5}$$

where η is the state vector, $e_v(k)$ is the instantaneous voltage regulation error vector, and matrix A_s has poles at specified frequencies to be canceled from $e_v(k)$—for example the fundamental, fifth, and seventh harmonics. All quantities are in the $\alpha\beta$ reference frame. The current command for the current controller can be obtained as

$$i_{ref}(k) = K_o X(k) + K_1\eta(k) \tag{6.6}$$

where the gains K_0 and K_1 are obtained by solving the linear quadratic optimization problem of the augmented system of X and η. Therefore, run-time gain calculation is not necessary.

6.4 REAL AND REACTIVE POWER CONTROL PROBLEMS

Since the DG unit uses a voltage source inverter with a strong voltage control, its output active and reactive power are determined by the unit's output voltage, including magnitude and phase angle, as stated in

$$P = \frac{V_{out}E}{X}\sin\,\delta, \tag{6.7}$$

$$Q = \frac{V_{out}^2 - V_{out}E\cos\delta}{X}, \tag{6.8}$$

where E is the equivalent main voltage, X is the equivalent line reactance where the resistance is ignored, and δ is the power angle. Since the DG unit output voltage control already exists, the task of the power controller is to generate voltage command

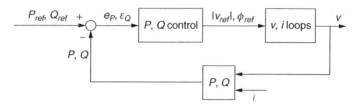

FIGURE 6.3 Control structure for grid-connected mode.

for the voltage controller based on the desired power values P_{ref} and Q_{ref} and actual values P and Q as illustrated in Fig. 6.3.

It is apparent that the desired DG output voltage V and the power angle δ can be calculated from Eqs. (6.7) and (6.8) given desired P and Q values and system parameter X. If this is true, the power control problem is solved. However, in practical systems, the above approach is not feasible based on the existing techniques due to the following three reasons:

- Equations (6.7) and (6.8) show that the power system parameters, both X and E, need to be known to solve the equations, which is difficult based on the existing approaches. Practically, the value of X may change due to the operation of the power system.
- Both are P and Q are sensitive to variation of X since it appears in the denominators and especially when X is small. The greater the difference between the power system capacity and the DG's power rating, the less the value of X could be.
- Both equations are nonlinear and are difficult to solve in real time, which prevents the idea being implemented in practice.

Therefore, power control solutions requiring knowledge of X have not been practically used, and people tend to search for other solutions (e.g., the integral control).

6.5 THE CONVENTIONAL INTEGRAL CONTROL

It can be observed from Eqs. (6.7) and (6.8) that both P and Q will be affected by only adjusting one of V and δ, which is the so-called coupling between P and Q. However, variations of V and δ have different levels of impact on P and Q as described in the following partial derivatives:

$$\frac{\partial P}{\partial \delta} = \frac{V_{out} E}{X} \cos \delta, \tag{6.9}$$

$$\frac{\partial P}{\partial V_{out}} = \frac{E}{X} \sin \delta, \tag{6.10}$$

$$\frac{\partial Q}{\partial \delta} = \frac{V_{out} E}{X} \sin \delta, \tag{6.11}$$

$$\frac{\partial Q}{\partial V_{out}} = \frac{2V_{out} - E \cos \delta}{X}. \tag{6.12}$$

These partial derivatives are plotted in a three-dimensional manner as shown in Fig. 6.4 to illustrate the significance of the impacts of V and δ variations on P and Q under different V and δ values. The values of V, E, and X are normalized in Fig. 6.4 for comparison purposes.

It can be observed from Fig. 6.4 that when $|\delta|$ is small and V is close to 1, which is true for large capacity power systems, $\partial P/\partial \delta$ is close to 1 and $\partial P/\partial V_{out}$ is close to

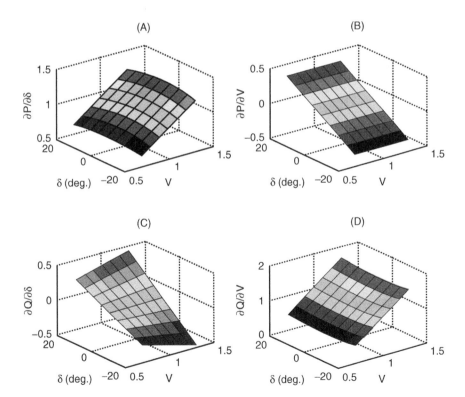

FIGURE 6.4 Sensitivity of P and Q to V and δ variations with normalized V, E, and X : (A)$\frac{\partial P}{\partial \delta}$, (B)$\frac{\partial P}{\partial V_{out}}$, (C)$\frac{\partial Q}{\partial \delta}$, (D)$\frac{\partial Q}{\partial V_{out}}$..

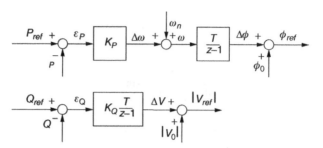

FIGURE 6.5 The power regulator for P and Q.

0 and, reversely, $\partial Q/\partial \delta$ is close to 0 and $\partial Q/\partial V_{out}$ is close to 1. This fact indicates that P is more sensitive to δ and Q is more sensitive to V_{out} especially when the DG unit is connected to a large capacity system where the power angle δ is usually small. The different levels of sensitivity of P and Q to δ and V_{out} provide a chance to control P and Q relatively independently, not completely independently though.

Based on the above analysis, an integral approach to conduct the power flow control can be developed to control P by adjusting δ and control Q by adjusting V_{out}. If the phase angle associated with the system voltage E is assumed to be 0, $\delta = \phi$ holds, where ϕ is the phase angle associated with V_{out}. The voltage and phase angle references can be generated as

$$\phi_{ref} = \int [K_P(P_{ref} - P) + \omega_n]\, dt + \phi_0, \tag{6.13}$$

$$V_{ref} = \int K_Q(Q_{ref} - Q)\, dt + V_0, \tag{6.14}$$

where $\omega n = 2\pi \times 60$ rad/sec is the system nominal angular frequency, and ϕ_0 and V_0 are the initial voltage and phase angle at the moment that the DG unit is connected to the grid from island running mode. The proposed power controller is illustrated in Fig. 6.5, where the integration is implemented in discrete time.

6.6 THE STABILITY PROBLEM

In the integral control, P and Q controls are decoupled under steady state due to the integration of the errors. However, in the transient, the P–Q coupling cannot be eliminated. Moreover, both P and Q are nonlinear functions of V_{out} and ϕ, which increases the complexity to analyze the system behavior. Due to the coupling issue and the nonlinearity, the stability of the power control must be investigated. Due to the strong regulation of the DG output voltage V_{out} and its phase angle ϕ, the dynamics of the voltage control loop can be simplified into a first-order system with a transfer function representation:

$$\phi(s) = \frac{a_\phi}{s + a_\phi} \phi_{ref}(s), \tag{6.15}$$

$$V_{out}(s) = \frac{a_V}{s + a_V} V_{ref}(s), \tag{6.16}$$

where a_ϕ and a_V are the inverses of the time constants of ϕ and V_{out} dynamics. Since the scope of this discussion is about power control stability of a DG unit connected to a large power system, whose time constant is in a range of seconds and is much greater than that of the voltage tracking response measured in a range of 0.01 sec or less, the DG power response has the room to be much slower than the voltage tracking and still fast compared to the power system. Therefore, it is reasonable to ignore the dynamics of the voltage tracking when power control is concerned, that is, $a_\phi \to \infty$, $a_V \to \infty$, $\phi \to \phi_{ref}$, and $V_{out} \to V_{ref}$

At the moment that the DG is switched from island mode to grid-connected mode, V_0 and ϕ_0 should match E and ϕE, where ϕE denotes the phase angle associated to E. Therefore, Eqs. (6.13) and (6.14) can be rewritten into

$$\delta = \int [K_P(P_{ref} - P) + \omega_n] \, dt, \tag{6.17}$$

$$\Delta V = \int K_Q(Q_{ref} - Q) \, dt, \tag{6.18}$$

where $\delta = \phi_{ref} - \phi E$ and $\Delta V = V_{ref} - E$. Assuming a large capacity power system with small power angle δ, it is reasonable to have $\sin \delta \approx \delta$ and $\cos \delta \approx 1$. Equations (6.17) and (6.18) can be rewritten in differential format:

$$\dot{\delta} = K_P \left(P_{ref} - \frac{E V_{out}}{X} \delta \right), \tag{6.19}$$

$$\Delta \dot{V} = K_Q \left(Q_{ref} - \frac{V_{out}}{X} \Delta V \right) \tag{6.20}$$

Since the dynamics of DG voltage tracking is ignored, the stability of the power loop can be evaluated using Lyapunov's direct method, where there is no external excitation, that is, $P_{ref} = 0$ and $Q_{ref} = 0$.

A Lyapunov function can be defined as

$$\xi(\Delta V, \delta) = \frac{1}{2} \Delta V^2 + \frac{1}{2} \delta^2, \tag{6.21}$$

where $\xi > 0$ holds unless $\Delta V = 0$ and $\delta = 0$. The derivative of the above function is

$$
\begin{aligned}
\dot{\xi}(\Delta V, \delta) &= \Delta V \cdot \Delta \dot{V} + \delta \cdot \dot{\delta} \\
&= -K_Q \frac{V_{out}}{X} \Delta V^2 - K_P \frac{E V_{out}}{X} \delta^2.
\end{aligned}
\tag{6.22}
$$

From Eqs. (6.22), it can be observed that $\dot{\xi}(\Delta V, \delta) < 0$ holds when $\Delta V \neq 0$ or $\delta \neq 0$, given positive values of KP, KQ, E, V_{out}, and X. Therefore, the proposed power control loop is asymptotically stable at the vicinity of the equilibrium point $\Delta V = 0$ and $\delta = 0$.

6.7 NEWTON–RAPHSON PARAMETER ESTIMATION AND FEEDFORWARD CONTROL NEWTON–RAPHSON PARAMETER IDENTIFICATION

The power flow control plant is governed by the two nonlinear equations given in Eqs. (6.7) and (6.8). Recall the concept mentioned in Section 6.4 that direct solution of V_{ref} and δ_{ref} is desired, assuming that the other parameters are known. This concept has not become practice only because it lacks means of implementation. The concept itself is very reasonable and inspiring. In these two equations, under conventional integral control, V, δ, P, and Q are all known. If the nonlinear equations can be solved for E and X, the Thevenin equivalent circuit parameters are obtained and can be used to improve the power flow control performance. Direct analytical solution of these equations is messy due to the nonlinearity. In this research, a real-time achievable numerical solution based on Newton–Raphson Method has been developed. The algorithm is presented as follows.

To avoid messy mathematical derivation, it is reasonable to replace the equations from reactance model based to susceptance model based; that is, use system susceptance B to take the place of $1/X$. This change does not affect the goal of identifying the system parameters to get achieved and meantime simplifies the problem and leads to a Newton–Raphson parameter estimator as depicted in Fig. 6.6. Rewrite the equations into the form of

$$
\begin{cases}
f(E,\ B) = BVE \sin \delta - P = 0, \\
g(E,\ B) = BV^2 - BVE \cos \delta - Q = 0.
\end{cases}
\tag{6.23}
$$

The Jacobian is then obtained as

$$
J_{est} =
\begin{bmatrix}
\dfrac{\partial f}{\partial E} & \dfrac{\partial f}{\partial B} \\[2ex]
\dfrac{\partial g}{\partial E} & \dfrac{\partial g}{\partial B}
\end{bmatrix},
\tag{6.24}
$$

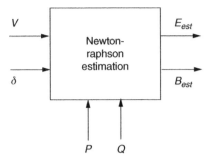

FIGURE 6.6 A Newton–Raphson Method-based nonlinear parameter estimator.

where

$$\frac{\partial f}{\partial E} = BV \sin \delta,$$

$$\frac{\partial f}{\partial B} = VE \sin \delta,$$

$$\frac{\partial g}{\partial E} = -BV \cos \delta,$$

and

$$\frac{\partial g}{\partial B} = V^2 - VE \, \cos \delta.$$

Given initial values E_0 and B_0, the iterations can be conducted. Solve the linearized equation

$$J_{est} \begin{bmatrix} \Delta E_k \\ \Delta B_k \end{bmatrix} = \begin{bmatrix} f(E_k, \ B_k) \\ g(E_k, \ B_k) \end{bmatrix} \tag{6.25}$$

for ΔE_k and ΔB_k, and

$$E_k + 1 = E_k - \Delta e_k, \tag{6.26}$$

$$B_k + 1 = B_k - \Delta B_k. \tag{6.27}$$

Newton–Raphson is known for fast convergence. However, if the nonlinear equation has saddles or multiple roots, the algorithm may not converge to the desired root. Therefore, the convergence condition needs to be checked before this approach can be practiced.

It can be observed from Eq. (6.24) that all four Jacobian elements hold monotonicity with given parameters. Therefore, no saddles or multiple roots exist and the iteration converges to the right solution.

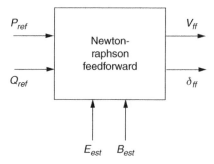

FIGURE 6.7   A Newton–Raphson Method-based nonlinear feedforward controller.

Due to the fast convergence, three to five iterations can yield solutions close enough to the true ones. Each control period can run one or multiple iterations, depending on the load of control tasks. Also, no derivatives need to be taken in generating the Jacobian, the linear equation solution can be obtained using a precalculated formula to save time, and the proposed parameter estimation technique can be implemented on DSP processor controllers in real time.

6.7.1   Newton–Raphson Feedforward Control

Given the real-time implementation of Newton–Raphson parameter estimation, the Thevenin equivalent circuit parameters of the power system can be approximately obtained. The acquired information can be used to perform P and Q feedforward control using the similar technique: With knowledge of the parameters E and B (from estimation) and P_{ref} and Q_{ref} (from desired power flow requirement), the voltage command can be solved also using the Newton–Raphson Method. This technique makes the idea of direct solution of V_{ref} and δ_{ref} implementable in real time and results in a P and Q feedforward controller as depicted in Fig. 6.7. The algorithm can be derived as follows. Rewrite the equations into the form of

$$f(V_{ref}, \delta_{ref}) = BV_{ref} E \sin \delta_{ref} - P_{ref} = 0,$$
$$g(V_{ref}, \delta_{ref}) = BV_{ref}^2 - BV_{ref} E \cos \delta_{ref} - Q_{ref} = 0. \tag{6.28}$$

The Jacobian is then obtained as

$$J_{ff} = \begin{bmatrix} \dfrac{\partial f}{\partial V_{ref}} & \dfrac{\partial f}{\partial \delta_{ref}} \\[2mm] \dfrac{\partial g}{\partial V_{ref}} & \dfrac{\partial g}{\partial \delta_{ref}} \end{bmatrix}, \tag{6.29}$$

where

$$\frac{\partial f}{\partial V_{ref}} = BE \sin \delta_{ref},$$

$$\frac{\partial f}{\partial \delta_{ref}} = BV_{ref} E \cos \delta_{ref},$$

$$\frac{\partial g}{\partial V_{ref}} = 2BV_{ref} - BE \cos \delta_{ref},$$

and

$$\frac{\partial g}{\partial \delta_{ref}} = BV_{ref} E \sin \delta_{ref}.$$

Given initial values $V_{ref,0}$ and $\delta_{ref,0}$, the iterations can be conducted. Solve the linearized equation

$$\begin{bmatrix} \Delta V_{ref} \\ \Delta \delta_{ref} \end{bmatrix} = [J_{ff}]^{-1} \begin{bmatrix} \Delta Q \\ \Delta P \end{bmatrix} \tag{6.30}$$

for $\Delta V_{ref,k}$ and $\Delta \delta_{ref,k}$, and

$$V_{ref,k+1} = V_{ref,k} - \Delta V_{ref,k}, \tag{6.31}$$

$$\delta_{ref,k+1} = \delta_{ref,k} - \Delta \delta_{ref,k}. \tag{6.32}$$

Similarly, the convergence condition needs to be checked before this approach can be practiced.

It can be observed from Eq. (6.29) that all four Jacobian elements hold monotonicity with given parameters. Therefore, no saddles or multiple roots exist and the iteration converges to the right solution.

Also similar to the parameter estimator proposed above, due to the fast convergence of Newton–Raphson Method, three to five iterations can yield solutions close enough to the true ones. Each control period can run one or multiple iterations, depending on the load of control tasks. Also, no derivatives need to be taken in generating the Jacobian, the linear equation solution can be obtained using a precalculated formula to save time, and the proposed parameter estimation technique can be implemented on microprocessor controllers in real time. Combining the proposed Newton–Raphson-based feedforward controller with the conventional integral control, it can be stated that even though small convergence errors still exist after a small number of iterations, the resulting V_{ref} and δ_{ref} are close to the steady-state value and stability of the integral control can be maintained due to the global stability of the technique proved in Section 6.6. The overall power regulator is shown in Fig. 6.8.

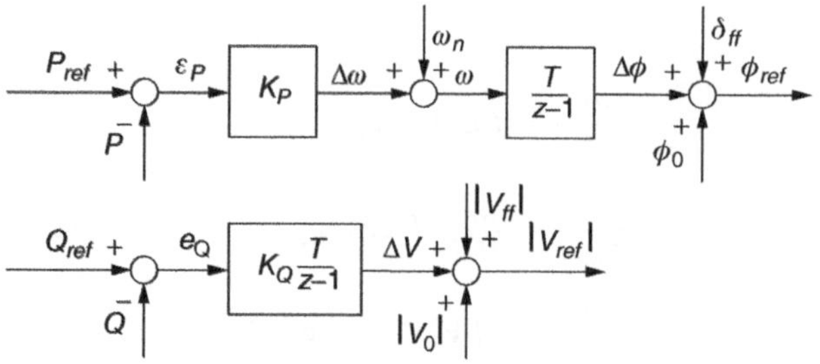

FIGURE 6.8 Power regulator combining integral control and feedforward.

6.8 HARMONIC POWER CONTROL

A harmonic power controller has been proposed to be applied together with the proposed fundamental power flow technique to handle harmonic distorted grid line voltage and prevent harmonic power from flowing between the DG unit and the utility grid. The block diagram of the ith harmonic power controller is shown in Fig. 6.9. This harmonic controller has the same algorithm of the conventional

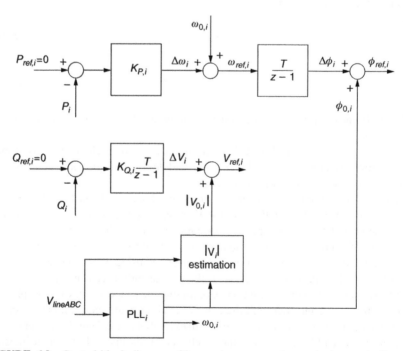

FIGURE 6.9 Control block diagram of harmonic compensation under harmonic distorted grid voltage.

Discrete 1-phase PLL

Pleme Glrauc Gilbert Sybille
Power System Laboratory, IREQ.
Hydro-Quebec

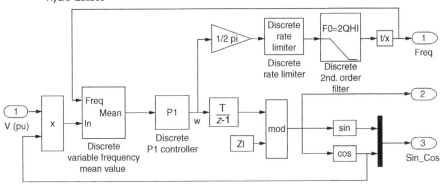

Note: sin(a)*cos(b) = 1/2*sin(a+b) + 1/2*sin(a−b)

FIGURE 6.10 Simulink® implementation of a PLL by Hydro-Quebec®.

integral power control of the fundamental power. In particular, $P_{ref,i}$ and $Q_{ref,i}$ are both zero for desired zero P_i and Q_i. The nominal frequency and phase angle of the ith harmonic, $\omega_{0,i}$ and $_{0,i}$, are obtained from a standard phase-locked loop (PLL) as given in Fig. 6.10.

The nominal voltage of the harmonic $|V_{0,i}|$ is calculated by a harmonic magnitude estimation algorithm as shown in Fig. 6.11.

6.9 SIMULATION RESULTS

Simulations have been conducted on a 5-kVA DG unit with a topology shown in Fig. 6.1 connected to a 120-V line-to-neutral power system with an equivalent reactance 0.1Ω under a number of difference scenarios as shown below.

6.9.1 In Island Mode

Under island mode, a 100% step load increase is applied to the DG output terminal. After steady state is reached, the load steps back down to zero. The voltage response under these transients are shown in Fig. 6.12.

Simulation data show that the voltage tracking error under steady state is nearly zero and the THD is 0.4%. It can be observed from Fig. 6.12 that the load disturbance has little impact on V_{out} waveform and the RMS transients last for only 0.02 sec with 3% or so peak variations. Recall that similar results obtained from the four-wire split DC bus topology have already been presented in previous chapters.

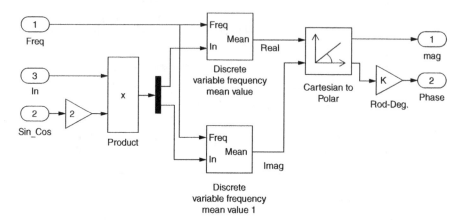

FIGURE 6.11   Simulink® implementation of a harmonic magnitude estimator by Hydro-Quebec®.

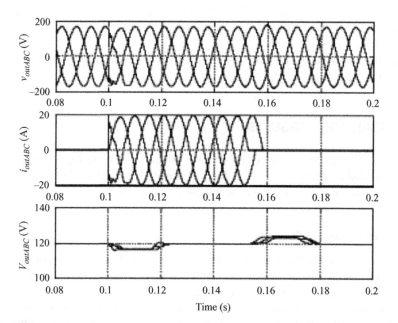

FIGURE 6.12   Transient response of V_{out} in instantaneous and RMS at step load increase from 0% to 100% and decrease from 100% to 0%.

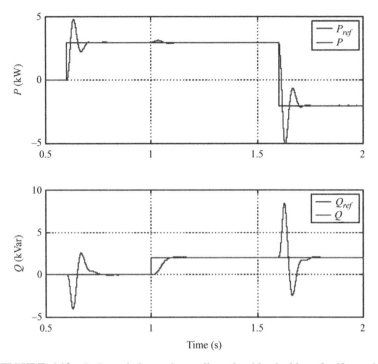

FIGURE 6.13 P, Q regulation under nonlinear local load without feedforward.

6.9.2 In Grid-Connected Mode

Under the grid-connected mode, P and Q output to the utility grid must be controlled for system stabilizing, compensation, or handling local load disturbances. Figure 6.13 illustrates the P and Q regulation under various step references under nonlinear local load without the Newton–Raphson parameter estimation and feedforward. Figure 6.14 illustrates it with the Newton–Raphson parameter estimation and feedforward. It can be observed from these figures that under the power command step changes, transient coupling between P and Q cannot be removed while the steady–state decoupling can be achieved. By comparing these two figures, it is obvious that the one with the Newton–Raphson parameter estimation and feedforward yields significant less transient coupling and overshoot and is therefore more effective.

When nonlinear local load exists, the line current i_{kine} is not supposed to be affected by the proposed control. Figure 6.15 exhibits the current waveforms at three different locations of the system including the line current i_{line}, the unit output current i_{out}, and the inverter current i_{inv}. The waveforms show that all current harmonics are taken by the DG unit and that the system line current is clean. This is because the voltage control loop eliminates the voltage harmonics at the DG output, which avoids harmonic current flow to or from the utility grid.

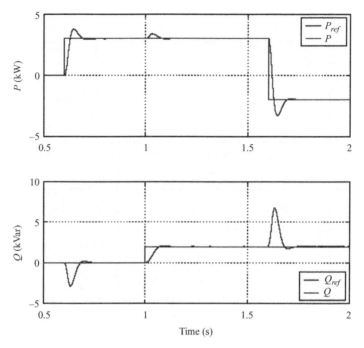

FIGURE 6.14 P, Q regulation under nonlinear local load with feedforward.

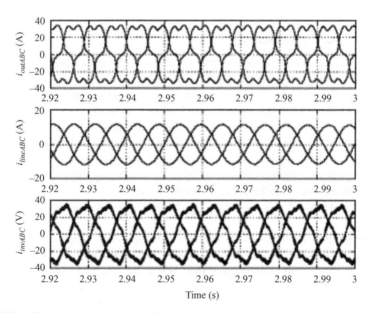

FIGURE 6.15 Current waveforms of DG unit output current i_{out}, system line current i_{line}, and inverter current i_{inv} under nonlinear local load.

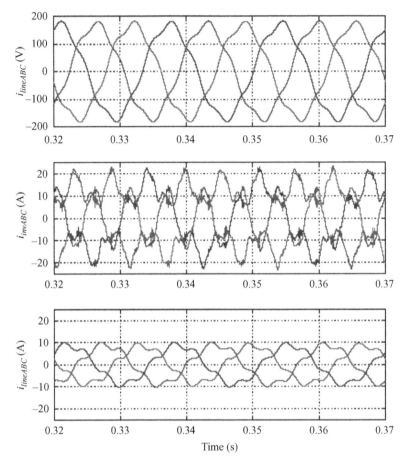

FIGURE 6.16 v_{line} *ABC*, i_{inv} *ABC*, and i_{line} *ABC* under the fifth harmonic distorted grid voltage without the fifth harmonic power control.

6.9.3 Line Current Conditioning Under Harmonic Distorted Grid Voltage

The effectiveness of the harmonic power controller has been demonstrated by comparing two simulation scenarios where one has while the other does not have the harmonic power controller under a fifth harmonic distorted utility grid voltage. The waveforms under the scenario where there is no harmonic power control is shown in Fig. 6.16 while the other one, where there is the harmonic power controller, is shown in Fig. 6.17.

It is clearly seen from trace 3 of the two figures that the result with the harmonic power control yields much less line current harmonics, which has verified the importance of this technique under harmonic corrupted utility grid voltage.

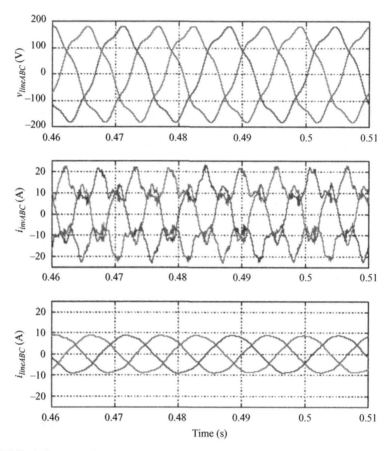

FIGURE 6.17 v_{line} ABC, i_{inv} ABC, and i_{line} ABC under the fifth harmonic distorted grid voltage with the fifth harmonic power control.

6.10 EXPERIMENTAL RESULTS

Power flow control experiments have been conducted on a 5-kVA grid-connectable inverter unit, where the main parameters are: DC bus voltage 540 V, output filter inductance 1.8 mH, filter capacitance 55 μF, Δ/Y isolation transformer equivalent leakage reactance 5%, transformer secondary filter capacitor 5 μF. SEMIKRON® IGBT modules SKM 50GB 123D are used in the inverter. A Texas Instruments TMS320LF2407A digital signal processor (DSP) with a Spectrum Digital® evaluation module (EVM) has been used as the digital controller. An Omron® relay controlled contactor is used to operate the grid connection.

The Newton–Raphson parameter estimation and feedforward control have been developed and included in the power control code on the DSP, where the estimated parameters are used in the feedforward. The power control performance of the real-time implementation of the technique has been demonstrated by Fig. 6.18 better

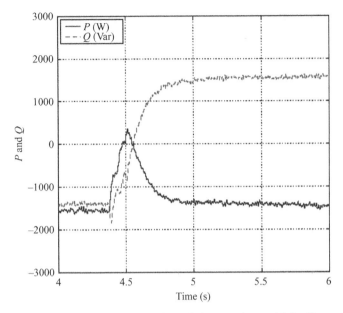

FIGURE 6.18 Experimental P, Q regulation transients with feedforward.

than the case where the technique is not used as shown in Fig. 6.19 in that the latter is much faster, even though the former has a slightly smaller but also slower transient coupling between P and Q.

The single-phase current transients under the above two scenarios have been presented and compared in Fig. 6.20, where the envelopes of the two current waveforms can be clearly seen and where the bottom trace, which includes the feedforward, yields better performance.

Since the harmonic power controller has not been implemented in experiments, the line current waveforms shown in Figs. 6.21 and 6.22 contain a significant amount of harmonic current. This is due to, of course, the harmonic distorted utility grid voltage, and the fundamental output voltage of the DG unit has been mostly canceled or evened out by that of the grid voltage, which makes the harmonic voltage components in the grid voltage take a more significant row in generating the line current. This explains why the line current contains much more harmonics compared to the load current.

This chapter has presented a power flow control approach for a single distributed generation unit connected to utility grid with a local load. The proposed control technique is based on a robust servomechanism voltage controller and a discrete-time sliding mode current controller designed for a three-phase three-wire inverter topology with isolation transformer. In order to obtain the parameters of the utility grid and use the information to generate feedforward control of power flow, a Newton–Raphson Method-based parameter estimation and feedforwrd control technique have been developed and combined with traditional integral power control.

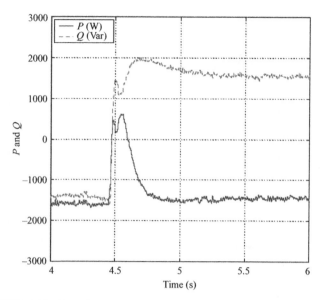

FIGURE 6.19 Experimental P, Q regulation transients without feedforward.

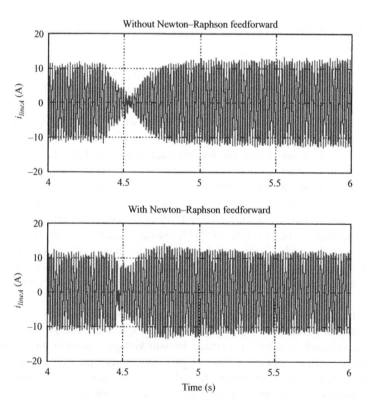

FIGURE 6.20 Experimental i_a transients without feedforward (**top trace**) and with feedforward (**bottom trace**).

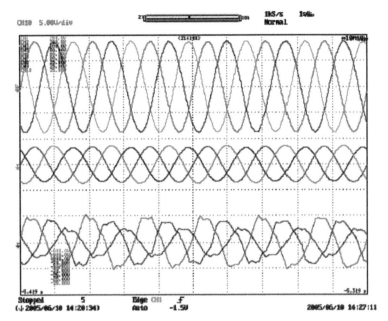

FIGURE 6.21 v_{line} *ABC* (**top**), i_{load} *ABC* (**middle**), and i_{line} *ABC* (**bottom**) under harmonic distorted grid voltage: $P = -1500$ W and $Q = -1500$ W.

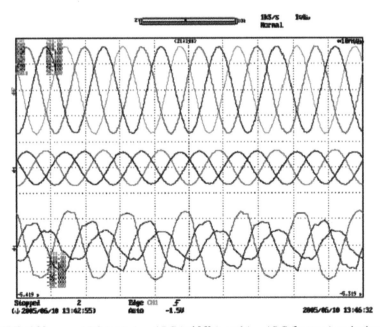

FIGURE 6.22 v_{line} *ABC* (top), i_{load} *ABC* (**middle**), and i_{line} *ABC* (**bottom**) under harmonic distorted grid voltage: $P = -1500$ W and $Q = 1500$ W.

The stability of the power control loop has been proved using the Lyapunov direct method. A harmonic power control technique based on PLL has been proposed to handle harmonic distorted utility grid voltage and yield harmonic free line current. Both simulation and experimental results under various scenarios have demonstrated the effectiveness of the proposed technique in power regulation and line current conditioning.

CHAPTER 7

ROBUST STABILITY ANALYSIS OF VOLTAGE AND CURRENT CONTROL FOR DISTRIBUTED GENERATION SYSTEMS

In this chapter, we will use the previous DG system consisting of DC power generated by green energy sources, an inverter, a transformer, and load as a standard project to present stability problem. The DG system, if operated in standalone mode, is typically used as on-site power. A state-space model of the system in DQ0 stationary reference is developed, and then the mathematical model is presented. A MATLAB simulation testbed is constructed to study the stability of the system.

7.1 INTRODUCTION

A feedback control system is said to achieve robust stability if it remains stable for all considered perturbations in the plant. The stability robustness of the system is evaluated by its tolerance to perturbations. In feedback-controlled PWM inverter systems—for example, an inverter-based three-phase distributed generation (DG) unit operated in standalone mode—load disturbance, noise, and parametric uncertainty of the electrical components in the circuit are the major plant perturbations that have significant impacts on both system stability and performance and therefore warrant detailed investigation. Robust-stability-related topics about PWM inverter-based systems have been addressed in the literature. We will present a state feedback control method of a PWM DC–DC converter for its robust stability under parametric uncertainty. This study checks whether the feedback system is stable by applying the Routh–Hurwitz stability tests but does not tell the stability margin or how stable

Integration of Green and Renewable Energy in Electric Power Systems. By A. Keyhani, M. N. Marwali, and M. Dai
Copyright © 2010 John Wiley & Sons, Inc.

the system is. A robust model reference adaptive control technique for uninterruptible power supplies should handle model inaccuracy. One approach is to use an H_∞ loop-shaping robust controller design technique with robust stability analysis. However, this technique does not perform well under nonlinear load, which significantly undermines its value for power supply applications. The DC–DC power converter is an analyzed controller using the structured singular value (μ) concept, which evaluates how stable the system is under the worst case of perturbation. This study uses admittance instead of resistance to model the DC load, which is proved convenient in the analysis. However, this design only considers load disturbances, and no parametric uncertainties are included in the perturbation. Another method is to develop a robust controller for a current source inverter (CSI). Both H_∞ loop-shaping and μ-analysis techniques are applied in approach, but no parametric uncertainty is considered which undermines the strength. Also, the robust controller design method for high-frequency resonant inverters can be modeled. This approach applies the H_∞ robust controller synthesis method provided in MATLAB® Robust Control Toolbox but only includes load and external input voltage in the perturbation.

A perfect robust servomechanism problem (RSP) controller guarantees exact asymptotic tracking of the fundamental frequency reference and error regulation of the load disturbance at each of the harmonic frequency included in the servocompensators. The perfect RSP guarantees this property independent of any perturbations in the plant as long as they do not destabilize system. The perfect RSP guarantees stability under a nominal plant without perturbation; however, the stability under perturbation is not guaranteed. Therefore, it is important to analyze the stability property of the controller under possible disturbances in order to ensure proper operation of the converter over its intended operating range.

In this chapter, the stability robustness of the system will be presented using *structured singular values* or a *μ-framework*. Specifically, perturbations due to load variations and parameter uncertainties of the system components are considered. A linear quadratic cost function with separate weighting scalars for plant states and servocompensator states have been used to find solutions to the perfect RSP. In this chapter, the stability robustness and transient response of the resulting control system will also be studied for different choices of these weighting scalars. The transient performance of the system is evaluated by performing moving window RMS calculations of the three-phase output voltages under transient load change from 0–100% resistive load.

7.2   THE STABILITY PROBLEM

We will first present a review of the voltages and current controller, followed by a summary of robust stability theory using structured singular value or μ-framework. The uncertainty model is then developed and used for verifying the robust stability of the system. We will use our standard model of DG system assuming that DC bus voltage is well-regulated as shown in Fig. 7.1. If the DC bus is not well-regulated, then the model of storage system should be included in the study.

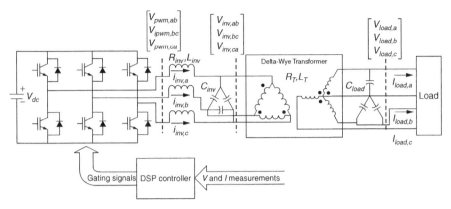

FIGURE 7.1   PWM inverter-based DG system in standalone mode.

7.2.1 Voltages and Current Control

Figure 7.1 shows a PWM inverter used for the DG system, where a constant DC voltage source is used to approximate most typical distributed generation sources, such as fuel cell, photovoltaic, wind, and micro-turbine generation systems. This approximation is surely reasonable when these sources work together with a stiff voltage regulation and secondary energy storage when necessary. The approximation is also reasonable even with an unregulated DC bus due to the adaptability of the PWM inverter as long as the DC voltage is high enough to generate the control command.

The DG system shown in Fig. 7.1, if operated in standalone mode, is typically used as on-site power or standby emergency power when the utility grid is not available or the utility power is accidentally lost due to fault.

A state-space model of the system in DQ0 stationary reference using per-unit notation can be developed and is given by Eqs. (7.1):

$$d\vec{v}_{inv,qd}/dt = \left(\vec{i}_{inv,qd} - \vec{i}_{snd',qd}\right)/c_{inv}, \tag{7.1a}$$

$$d\vec{i}_{inv,qd}/dt = \left(-r_{inv} \cdot \vec{i}_{inv,qd} - \vec{v}_{inv,qd} + \vec{v}_{pwm,qd}\right)/l_{inv}, \tag{7.1b}$$

$$d\vec{v}_{load',qd}/dt = \left(\vec{i}_{snd',qd} - \vec{i}_{load',qd}\right)/c_{load}, \tag{7.1c}$$

$$d\vec{i}_{snd',qd}/dt = \left(-r_T \cdot \vec{i}_{snd',qd} + \vec{v}_{inv,qd} - \vec{v}_{load',qd}\right)/l_T, \tag{7.1d}$$

$$dv_{load,0}/dt = \left(i_{snd,0} - i_{load,0}\right)/c_{load}, \tag{7.1e}$$

$$di_{snd,0}/dt = \left(-r_T \cdot i_{snd,0} - v_{load,0}\right)/c_{load}, \tag{7.1f}$$

where the following per-unit capacitances, inductances, and resistances are defined from their corresponding per-unit values:

$$c_{inv} \equiv 1/(\omega_f \cdot xc_{inv}), \quad c_{load} \equiv 1/(\omega_f \cdot xc_{load}), \quad l_{inv} \equiv xl_{inv}/\omega_f,$$

$$l_{trans} \equiv xl_T/\omega_f, \quad r_{inv} \equiv xr_{inv}, \quad r_T \equiv xr_T$$

The per-unit values of the capacitances, inductances, and resistances are calculated from

$$xl_{inv} \equiv \omega_f L_{inv}/Z_1, \quad xr_{inv} \equiv R_{inv}/Z_1, \quad xc_{inv} \equiv 1/(\omega_f 3C_{inv})/Z_1,$$

$$xl_T \equiv \omega_f L_T/Z_2, \quad xr_T \equiv R_T/Z_2, \quad xc_{load} \equiv 1/(\omega_f C_{load})/Z_2,$$

where the base impedances Z_1 and Z_2 are defined as

$$Z_1 \equiv V_1/I_1, \qquad Z_2 \equiv V_2/I_2,$$

where V_1 and V_2 denote the rated line-to-neutral primary and secondary transformer voltages, and I_1 and I_2 denote the rated primary and secondary currents, respectively. The voltages and currents in Eqs. (7.1) are the DQ0 stationary reference frame variables of the per-unit voltages and currents in ABC given as

$$\vec{v}_{pwm,abc} = \vec{v}_{pwm,abc}/V_1, \qquad \vec{v}_{inv,abc} = \vec{v}_{inv,abc}/V_1$$

$$\vec{i}_{inv,abc} = \vec{i}_{inv,abc}/I_1, \qquad \vec{v}_{load,abc} = \vec{v}_{load,abc}/V_2$$

$$\vec{i}_{load,abc} = \vec{i}_{load,abc}/I_1$$

where the actual *ABC* voltages and currents are defined as follows:

$$\vec{v}_{pwm,abc} = \begin{bmatrix} V_{pwm,ab} & V_{pwm,bc} & V_{pwm,ca} \end{bmatrix}^T/\sqrt{3}$$

$$\vec{v}_{inv,abc} = \begin{bmatrix} V_{inv,ab} & V_{inv,bc} & V_{inv,ca} \end{bmatrix}^T/\sqrt{3}$$

$$\vec{v}_{load,abc} = \begin{bmatrix} V_{load,a} & V_{load,b} & V_{load,c} \end{bmatrix}^T,$$

$$\vec{i}_{load,abc} = \begin{bmatrix} I_{load,a} & I_{load,b} & I_{load,c} \end{bmatrix}^T$$

$$\vec{i}_{snd,abc} = \begin{bmatrix} I_{snd,a} & I_{snd,b} & I_{snd,c} \end{bmatrix}^T,$$

$$\vec{i}_{inv,abc} = \begin{bmatrix} I_{inv,a} - I_{inv,b}, & I_{inv,b} - I_{inv,c}, & I_{inv,c} - I_{inv,a} \end{bmatrix}^T/\sqrt{3}$$

The following changes of variables have been used for the secondary transformer currents, load currents, and load voltages in Eqs. (7.1):

$$\vec{i}_{snd',qd} = tri_{qd} \cdot \vec{i}_{snd,qd} \tag{7.2a}$$

$$\vec{i}_{load',qd} = tri_{qd} \cdot \vec{i}_{load,qd} \tag{7.2b}$$

$$\vec{v}_{load',qd} = trv_{qd}^{-1} \cdot \vec{v}_{load,qd} \tag{7.2c}$$

with matrices tri_{qd} and trv_{qd} defined as

$$tri_{qd} = \left[K_s \cdot tr_i \cdot K_s^{-1} \right]_{col1,2}^{row1,2} = \begin{bmatrix} -1 & 0 \\ 0 & -1 \end{bmatrix}, \tag{7.3a}$$

$$trv_{qd} = \left[K_s \cdot tr_v \cdot K_s^{-1} \right]_{col1,2}^{row1,2} = \begin{bmatrix} -1 & 0 \\ 0 & -1 \end{bmatrix}. \tag{7.3b}$$

Matrices tr_i and tr_v in Eqs. (7.3) denote currents and voltages transformations of a particular delta-wye transformer—for example,

$$tr_i = \frac{1}{3} \begin{bmatrix} -2 & 1 & 1 \\ 1 & -2 & 1 \\ 1 & 1 & -2 \end{bmatrix}, \quad tr_v = \begin{bmatrix} -1 & 0 & 0 \\ 0 & -1 & 0 \\ 0 & 0 & -1 \end{bmatrix}.$$

From Eq. (7.1) it can be seen that the zero variables are not affected by the control inputs and therefore need not be considered in the controller design. Moreover, the D and Q axes are completely decoupled and have the same dynamics. Therefore, a controller can be developed and analyzed for one of the dq axis, and the same controller can be used for the other.

Figure 7.2 shows the voltages and currents control. An RSP controller is used for the voltage control and a discrete-time sliding mode (DSM) controller is used for the current control.

The DSM controller is used in the current loop to limit the inverter current under overload condition because of the fast and no overshoot response it provides. Suppose a discrete form of the L–C filter dynamics given in Eq. (7.1a) and (7.1b) are

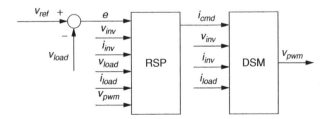

FIGURE 7.2   The RSP and DSM controllers.

given by

$$\vec{x}_1 (k + 1) = A_1^* \vec{x}_1 (k) + B_1^* \vec{u} (k) + E_1^* \vec{d}_1 (k) \qquad (7.4)$$

where the states are $\vec{x}_1 = \left[\vec{v}_{inv}, \vec{i}_{inv} \right]$, the inputs $u = v_{pwm}$, and disturbances $\vec{d}_1 = i_{snd}$. Note that the subscript dq has been dropped since the controller is designed for one axis but apply for both. To force the inverter currents to follow their commands, the sliding mode surface is chosen as $s (k) = C_1 \cdot \vec{x}_1 (k + 1) - i_{cmd} (k)$, where $C_1 \cdot \vec{x}_1 (k + 1) = i_{inv} (k + 1)$, so that when discrete sliding mode occurs, we have $\vec{s} (k) = 0$ or $i_{inv} (k + 1) = i_{cmd} (k)$. The existence of the discrete sliding mode can then be guaranteed if the control is given by

$$v_{pwm}(k) = \left(C_1 B_1^* \right)^{-1} \left(\vec{i}_{cmd} - C_1 A_1^* \vec{x}_1 (k) - C_1 E_1^* \vec{d}_1 (k) \right). \qquad (7.5)$$

The RSP is adopted for voltage control due to its capability to perform zero steady-state tracking error under unknown load and eliminate harmonics of any specified frequencies with guaranteed system stability for nominal plant parameters and load variations. The theory behind the RSP is based on the solution of robust servomechanism problem where the internal model principle and the optimal control theory for linear systems are combined. The RSP controller consists of a discrete form of the continuous servocompensator:

$$\dot{\vec{\eta}} = A_c \vec{\eta} + B_c e_V, \quad e_V = V_{ref} - V_{load}, \qquad (7.6)$$

where

$$\vec{\eta} = \left[\vec{\eta}_1, \vec{\eta}_2, \dots, \vec{\eta}_n \right]^T \quad \vec{\eta}_i \in R^2, \quad i = 1, 2, \dots, n,$$

$$A_c = \text{block diag} \left[Ac_1, Ac_2, \dots, Ac_n \right],$$

$$B_c = \left[Bc_1, Bc_2, \dots, Bc_n \right]^T$$

with

$$Ac_i = \begin{pmatrix} 0 & 1 \\ -\omega_i^2 & 0 \end{pmatrix}, \quad i = 1, 2, \dots, n,$$

$$Bc_i = \begin{pmatrix} 0 & 1 \end{pmatrix}^T, \quad i = 1, 2, \dots n.$$

Note that each of the blocks $\dot{\vec{\eta}}_i = Ac_1 \vec{\eta}_i + Bc_i \vec{e}_{vqd}$ represents a state-space implementation of the continuous transfer function $1/(z^2 + \omega_i^2)$, where each $\omega_i = 2\pi f_i$ represents the fundamental frequency to track and each of the harmonic frequencies to be eliminated. For a 60-Hz DG system with desire to eliminate fifth and seventh harmonics, for example, we use $\omega_1 = 2\pi \cdot 60$, $\omega_2 = 2\pi \cdot 5 \cdot 60$, and $\omega_3 = 2\pi \cdot 7 \cdot 60$.

To design the *voltage controller* using the perfect RSP, we need to consider a combination of the true plant in Eq. (7.1) and the discrete time sliding mode current controller (7.5) as the equivalent "plant" seen by the outer voltage loop. Assuming a discrete form of Eq. (7.1) given by $\vec{x}_p^*(k+1) = A_p^* \vec{x}_p^*(k) + B_p^* v_{vpwm}(k)$, the augmented true plant and discrete sliding mode current controller can be found as in

$$\vec{x}_p^*(k+1) = A_d \vec{x}_p^*(k) + B_d \cdot i_{cmd}(k) \tag{7.7}$$

with

$$A_d = A_p^* - B_p^* \left(C_1 B_1^*\right)^{-1} \left(B_1^* C_{11} + E_1^* C_{12}\right),$$

$$B_d = B_p^* \left(C_1 B_1^*\right)^{-1},$$

$$C_{11} = \begin{bmatrix} 1 & 0 & 0 & 0 & 0 \\ 0 & 1 & 0 & 0 & 0 \end{bmatrix},$$

$$C_{12} = \begin{bmatrix} 0 & 0 & 0 & 1 & 0 \end{bmatrix}.$$

The complete perfect RSP voltage controller is given by states feedback:

$$i_{cmd}(k) = K_0 x_p^*(k) + K_1 \eta(k), \tag{7.8}$$

where the gains $K = \begin{bmatrix} K_0 & K_1 \end{bmatrix}$ are found by minimizing a certain linear quadratic cost function for the augmented "equivalent plant" (7.7) and a discrete form of the servocompensator (7.6):

$$\begin{bmatrix} \vec{x}_p^*(k+1) \\ \eta(k+1) \end{bmatrix} = \begin{bmatrix} A_d & 0 \\ -B_c^* C & A_c^* \end{bmatrix} \begin{bmatrix} \vec{x}_p^*(k) \\ \eta(k) \end{bmatrix} + \begin{bmatrix} B_d \\ -B_c^* D \end{bmatrix} u_1(k). \tag{7.9}$$

To achieve desired transient performance and guarantee robust stability under plant uncertainties, the following linear quadratic cost function has been used in this chapter:

$$J_\varepsilon = \sum_{k=0}^{\infty} \left(\begin{array}{l} w_p \cdot x_p^*(k)' \cdot x_p^*(k) + w_{S1} \cdot \eta_1(k)' \cdot \eta_1(k) \\ + w_{SH} \sum_h \eta_h(k)' \cdot \eta_h(k) + \varepsilon \cdot u(k)' u(k) \end{array} \right), \tag{7.10}$$

where w_p, w_S and w_{SH} represent weighting scalars for plant states (x_p^*), fundamental servocompensator states (η_1), and harmonics servocompensator states (η_h). Solution to this linear quadratic optimization problem is well known and can be found using MATLAB command *dlqr*.

The effectiveness of the technique can be demonstrated by experimental results shown in Fig. 7.3, which exhibits well-regulated sinusoidal output voltage waveforms under various types of load.

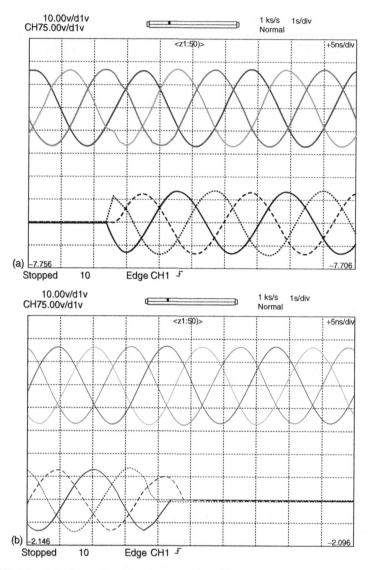

FIGURE 7.3   Experimental results of the standalone DG system under different scenarios, where the top traces are three-phase output voltages and the bottom traces are three-phase load currents. **(a)** Transient response at load stepping up from 0% to 100%. **(b)** Transient response at load stepping down from 100% to 0%. **(c)** Steady state-performance under three-phase nonlinear load.

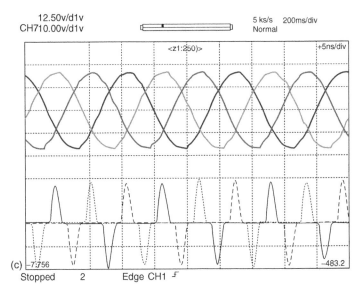

FIGURE 7.3 *(Continued)*

7.3 ROBUST STABILITY ANALYSIS USING STRUCTURED SINGULAR VALUE μ

Structured singular value μ can be used to analyze and evaluate the stability robustness of a multi-input–multi-output (MIMO) linear system under structured perturbations. In order to use the μ-*framework* to analyze robust stability of a linear system under perturbation, the problem needs to be recast into a feedback loop diagram of Fig. 7.3 where M represents a known stable MIMO transfer function of the linear system with n inputs and n outputs and Δ represents a structured uncertainty matrix of the form (7.11):

$$\Delta = \left\{ diag\left[\delta_1 I_{r1}, \ldots \delta_s I_{rs}, \Delta_1 \ldots \Delta_F\right] : \delta_i \in C, \Delta_j \in C^{m_j \times m_j} \right\}, \qquad (7.11)$$

where $\sum_{i=1}^{S} r_i + \sum_{j=1}^{F} m_j = n.$

The structured singular value of M with respect to the uncertainty set Δ is defined as

$$\mu_\Delta(M) = \frac{1}{\min\left\{\bar{\sigma}(\Delta) : \Delta \in \Delta, \det(I - M\Delta) = 0\right\}}.$$

The Generalized Small-Gain Theorem provides robust stability result of the system using the structured singular value. It states that, if nominal $M(s)$ is stable then the

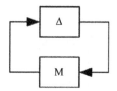

FIGURE 7.4   Representation of a linear system with uncertainties.

perturbed system $(I - M\Delta)^{-1}$ is stable for all stable Δ_i for which $\|\Delta_i\|_\infty \le 1$ if and only if $\mu_\Delta (M (j\omega)) < 1$ for all $\omega \in R$.

7.3.1 Uncertain Open-Loop Model

In order to use the μ-framework to analyze robust stability, the problem needs to be recast to that of Fig. 7.4. A class of general feedback loops called *linear fractional transformations* (LFT) can be used to achieve this. Using the state-space model (7.1), a single-phase equivalent circuit of the converter with RL load can be derived as shown in Fig. 7.5. This represents the open-loop model of the plant.

The dynamic equations of the plant are given by

$$dv_{inv}/dt = (i_{inv} - \cdot i_{snd'})/c_{inv}, \tag{7.12a}$$

$$di_{inv}/dt = (-r_{inv} \cdot i_{inv} - v_{inv} + v_{pwm})/l_{inv}, \tag{7.12b}$$

$$dv_{load}/dt = (isnd_{snd} - g_{load}v_{load'} - i_L)/c_{load}, \tag{7.12c}$$

$$di_{snd'}/dt = (-r_T \cdot i_{snd'} + v_{inv} - v_{load'})/l_T, \tag{7.12d}$$

$$di_L/dt = \lambda_{load}v_{load'}, \tag{7.12e}$$

$$i'_{load} = g_{load}v'_{load} + i_L, \tag{7.12f}$$

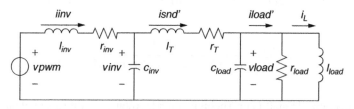

FIGURE 7.5   Open-loop model of the nominal plant.

where $g_{load} = 1/r_{load}$ is the per-unit conductance of the load and $\lambda_{load} = 1/l_{load}$ is the inverse of the per-unit load inductance.

Let us assume that the following parameter variations exist in the system due to manufacturing tolerances of the components used and/or errors in the parameters identification processes:

- Inverter filter capacitor tolerance: $\pm 6\%$
- Inverter filter inductor tolerance: $\pm 15\%$
- Inverter filter inductor losses tolerance: $\pm 50\%$
- Output filter capacitor tolerance: $\pm 6\%$
- Transformer filter inductor tolerance: $\pm 15\%$
- Transformer filter inductor losses tolerance: $\pm 50\%$

Furthermore, the specification of the unit requires that it operate stably without degradation in performance for load from 0% to 200% with power factor of 0.8 lagging at maximum load. This specification represents variation of the resistive load from 0% to 7.160%, and inductive load from 0% to 120%.

The above parameter variations due to manufacturing tolerances and load variations can be precisely written as follows:

$$c_{inv} = c_{inv0}\left(1 + \tau_{cinv} \cdot \delta_{cinv}\right), \quad |\delta_{cinv}| < 1, \tag{7.13a}$$

$$l_{inv} = l_{inv0}\left(1 + \tau_{linv} \cdot \delta_{linv}\right), \quad |\delta_{linv}| < 1, \tag{7.13b}$$

$$c_{load} = c_{load0}\left(1 + \tau_{cload} \cdot \delta_{cload}\right), \quad |\delta_{cload}| < 1, \tag{7.13c}$$

$$l_T = l_{T0}\left(1 + \tau_{lT} \cdot \delta_{lT}\right), \quad |\delta_{lT}| < 1, \tag{7.13d}$$

$$r_{inv} = r_{inv0}\left(1 + \tau_{rinv} \cdot \delta_{rinv}\right), \quad |\delta_{rinv}| < 1, \tag{7.13e}$$

$$r_T = r_{T0}\left(1 + \tau_{rT} \cdot \delta_{rT}\right), \quad |\delta_{rT}| < 1, \tag{7.13f}$$

$$g_{load} = g_{load0}\left(1 + \tau_{gload} \cdot \delta_{gload}\right), \quad |\delta_{gload}| < 1, \tag{7.13g}$$

$$\lambda_{load} = \lambda_{load0}\left(1 + \tau_{\lambda load} \cdot \delta_{\lambda load}\right), \quad |\delta_{\lambda load}| < 1, \tag{7.13h}$$

where the terms with subscript zero indicate the nominal values of the parameters and τ terms denote their percentage tolerances given above. The following load parameters can be used to represent the desired load variation above: $g_{load0} = 0.8$, $\lambda_{load0} = 0.6\omega_f$, with $\tau_{gload} = \tau_{\lambda load} = 1.0$.

Parameters c_{inv}, l_{inv}, c_{load}, and l_T appear in denominators in Eqs. (7.12) and can be represented using lower LFT similar to that shown in Fig. 7.6a for $1/c_{inv}$. LFTs for

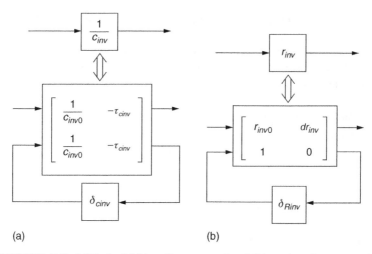

FIGURE 7.6   LFTs for (**a**) $1/c_{inv}$ (inverse term) and (**b**) r_{inv} (non-inverse term).

r_{inv}, r_T, r_{load}, and λ_{load} are constructed similar to that of Fig. 7.6b for r_{inv}, where

$$dr_{inv} = r_{inv0} \cdot \tau_{rinv}, \tag{7.14a}$$

$$dr_T = r_{T0} \cdot \tau_{rT}, \tag{7.14b}$$

$$dg_{load} = g_{load0} \cdot \tau_{gload}, \tag{7.14c}$$

$$d\lambda_{load} = \lambda_{load0} \cdot \tau_{\lambda load}, \tag{7.14d}$$

For each uncertain term, the corresponding LFT as in Fig. 7.6 can be substituted into the dynamic model of Eq. (7.12) as shown in Fig. 7.7 with the uncertain perturbation (δ) blocks separately represented by $\Delta(s)$ as shown in Fig. 7.8, where $\Delta(s)$ is given by $\{\delta_{C_{inv}}, \delta_{L_{inv}}, \delta_{C_{load}}, \delta_{LT}, \delta_{R_{inv}}, \delta_{RT}, \delta_{G_{load}}, \delta_{G_{load}}\}$. The block $P(s)$ in Fig. 7.8 contains all the known model information of the plant and will be referred to as the nominal open-loop plant model. The combination of $P(s)$ and $\Delta(s)$ forms the uncertain open-loop model.

The state-space model of the nominal plant P can be derived by inspection of Fig. 7.7 with the following states, inputs, and outputs variables:

States: $X = \begin{bmatrix} v_{inv} & i_{inv} & v_{load} & i_{snd} & i_L \end{bmatrix}^T$.

Inputs: $U = \begin{bmatrix} v_{pwm} & W^T \end{bmatrix}^T$, where

$$W = \begin{bmatrix} w_{cinv} & w_{linv} & w_{cload} & w_{lT} & w_{rinv} & w_{rT} & w_{gload} & w_{\lambda load} \end{bmatrix}^T.$$

Outputs: $Y = \begin{bmatrix} v_{inv} & i_{inv} & v_{load} & i_{load} & Z^T \end{bmatrix}^T$, where

$$Z = \begin{bmatrix} z_{cinv} & z_{linv} & z_{cload} & z_{lT} & z_{rinv} & z_{rT} & z_{gload} & z_{\lambda load} \end{bmatrix}^T.$$

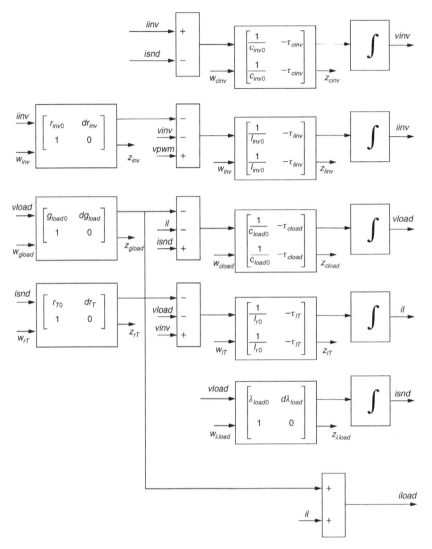

FIGURE 7.7 Block diagram of the nominal plant $P(s)$ with LFTs of the uncertain parameters.

The state-space equations are given as

$$\dot{X} = A_P X + B_P U,$$
$$Y = C_P X + D_P U,$$

(7.15)

where

$$A_P = [A_{nom}], \quad B_P = \begin{bmatrix} B_{nom} & B_{del} \end{bmatrix}, \quad C_P = \begin{bmatrix} C_{nom} \\ C_{del} \end{bmatrix}, \quad D_P = \begin{bmatrix} D_{nom} \\ D_{del} \end{bmatrix}.$$

The terms with subscription *nom* are the nominal model parameters derived from Eq. (7.12) and the ones with subscription *del* are defined in Eqs. (7.16), (7.17), and (7.18), respectively.

7.3.2 Uncertain Closed-Loop Model

In order to use the μ-framework to analyze the robust stability of the system, the system needs to be recast in to that of Fig. 7.4. In this case, the system M comprises of the nominal open-loop plant $P(s)$ and the controller loop closed around it as illustrated in Fig. 7.8 where the vectors X, W, and Z as defined above. MATLAB with its Control System Toolbox and μ-Analysis and Synthesis Toolbox has been utilized to achieve this purpose.

Since the controller is implemented in the discrete time system, the following steps are necessary to obtain the closed-loop plant model:

1. Apply a zero-order hold transformation to the continuous plant P, to include the effect of the sample and hold process of the digital sampling process.
2. Transform the discretized plant back to continuous system in the *w-plane* by applying an inverse Tustin transformation. This transformation has the property of preserving the frequency response of the discrete time systems
3. Obtain state-space representations of the *RSP* and Discrete Sliding Mode controller with inputs and outputs definitions as shown in Fig. 7.2. The combined controller state-space system can be calculated using *sysic* (μ -Toolbox) command in MATLAB.
4. The combined controller system is then transformed into the *w-plane* in continuous domain using the inverse Tustin transformation
5. Finally, the closed-loop plant model M is obtained by invoking the MATLAB *sysic* command for the transformed nominal plant model from step 2 and the controller system obtained in step 4.

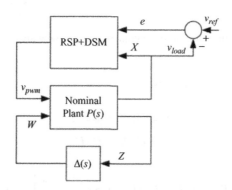

FIGURE 7.8   Uncertain closed-loop model.

$$
B_{del} =
\begin{bmatrix}
-\tau_{cinv} & 0 & 0 & 0 & 0 & 0 & 0 & 0 \\
0 & -\tau_{linv} & 0 & 0 & \dfrac{-dr_{inv}}{l_{inv0}} & 0 & 0 & 0 \\
0 & 0 & -\tau_{cload} & 0 & 0 & 0 & \dfrac{-dg_{load}}{c_{load0}} & 0 \\
0 & 0 & 0 & -\tau_{lT} & 0 & \dfrac{-dr_T}{l_{T0}} & 0 & 0 \\
0 & 0 & 0 & 0 & 0 & 0 & 0 & d\lambda_{loadl}
\end{bmatrix},
$$

$$(7.16)$$

$$
C_{del} =
\begin{bmatrix}
0 & \dfrac{1}{c_{inv0}} & 0 & \dfrac{-1}{c_{inv0}} & 0 \\
\dfrac{-1}{l_{inv0}} & \dfrac{-r_{inv0}}{l_{inv0}} & 0 & 0 & 0 \\
0 & 0 & \dfrac{-g_{load0}}{c_{load0}} & \dfrac{1}{c_{load0}} & \dfrac{-1}{c_{load0}} \\
\dfrac{1}{l_{T0}} & 0 & \dfrac{-1}{r_{T0}} & \dfrac{-r_{T0}}{r_{T0}} & 0 \\
0 & 1 & 0 & 0 & 0 \\
0 & 0 & 0 & 1 & 0 \\
0 & 0 & 1 & 0 & 0 \\
0 & 0 & 1 & 0 & 0
\end{bmatrix},
$$

$$(7.17)$$

$$
D_{del} =
\begin{bmatrix}
0 & -\tau_{cinv} & 0 & 0 & 0 & 0 & 0 & 0 & 0 \\
\dfrac{1}{l_{inv0}} & 0 & -\tau_{linv} & 0 & 0 & \dfrac{-dr_{inv}}{l_{inv0}} & 0 & 0 & 0 \\
0 & 0 & 0 & -\tau_{cload} & 0 & 0 & \dfrac{-dg_{load}}{c_{load0}} & 0 & 0 \\
0 & 0 & 0 & 0 & -\tau_{lT} & \dfrac{-dr_T}{l_{T0}} & 0 & 0 & 0 \\
0 & 0 & 0 & 0 & 0 & 0 & 0 & 0 & 0 \\
0 & 0 & 0 & 0 & 0 & 0 & 0 & 0 & 0 \\
0 & 0 & 0 & 0 & 0 & 0 & 0 & 0 & 0 \\
0 & 0 & 0 & 0 & 0 & 0 & 0 & 0 & 0
\end{bmatrix}
$$

$$(7.18)$$

7.4 TUNING THE CONTROLLER PERFORMANCE

Recall earlier that the w_p, w_S, and w_{SH} in cost function (7.10) are the weighting scalars for the plant states, fundamental state, and harmonic servocompensator states, respectively. The selection of values of these scalars provides a way of tuning the controller for desired transient performance and stability robustness. The structured singular value discussed in the previous sections will be used to evaluate the robust stability of each controller resulting from different choices of weighting scalars. A time response simulation of the single-phase equivalent circuit will be used to compare the transient performance of each controller. To quantify the transient performance, it is common in the industry to use the deviation of the RMS output voltage from its nominal value during a 100% resistive load change as a performance measure. A less than 5% deviation of output voltage under 100% resistive load transient is not an uncommon specification in the industry for a high-performance DG unit.

Consider the following cases of weighting scalars:

Case 1: $w_p = 0.5$, $w_s = 5 \times 10^5$, and $w_{SH} = w_S$
Case 2: $w_p = 0.1$, $w_s = 5 \times 10^5$, and $w_{SH} = w_S$
Case 3: $w_p = 0.05$, $w_s = 5 \times 10^5$, and $w_{SH} = w_S$
Case 4: $w_p = 0.005$, $w_s = 5 \times 10^5$, and $w_{SH} = w_S$
Case 5: $w_p = 0.0001$, $w_s = 5 \times 10^5$, and $w_{SH} = w_S$

In all the cases above, equal weighting scalars are applied to the fundamental compensator states and the harmonic states ($w_{SH} = w_S$), while the plant states weighting w_p in each case is decreased from 0.5 all the way down to 0.0001. The time response simulations for Case 1 is illustrated in Fig. 7.9 showing the output voltage, reference voltage, load current, and RMS variation of the output voltage during both 0 to 100% and 100% to 0% resistive load transients. From Fig. 7.9 it can be seen that the output voltage RMS deviates as much as close to 20% for Case 1.

The time response similar to Fig. 7.9 was obtained for each of the cases 1 to 5 and the resulting RMS output voltage variations are plotted collectively in Fig. 7.10. It can be seen that as the scalar weighting w_p decreases, the transient performance improves, with the RMS variation as little as 2% for the case $w_p = 0.0001$. These results are not unexpected since the weighting scalars represent the penalty applied to each state in the system. Intuitively, decreasing w_p while keeping the compensator states weighting the same, decreases the penalty cost applied to the plant states in the cost function to be minimized. This results in allowing the plant states to move more freely and hence faster response. Notice, however, that only Case 4 and Case 5 results in RMS variations of less than 5%.

To analyze the robust stability, the upper bound of the structured singular value in each case is plotted in Fig. 7.11. It can be seen that only Case 1, Case 2, and Case 3 achieve robust stability under the considered structured perturbations, with the peak value of $\mu_\Delta[M(j\omega)]$ in each case being less than 1. Case 4 and Case 5—the only

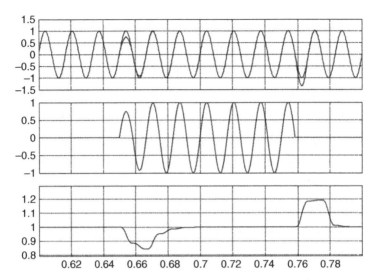

FIGURE 7.9 Transient response for Case 1: $w_p = 0.5$, $w_s = 5 \times 10^5$, and $w_{SH} = w_S$. The top trace is the output voltage and its reference, the middle trace is the load current, and the bottom trace is RMS variations of the output voltage, all versus time (second).

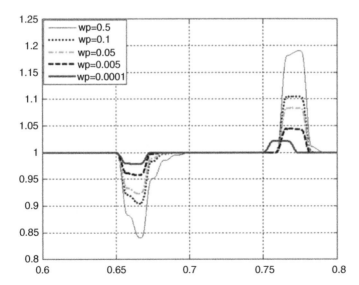

FIGURE 7.10 RMS output voltage variations during 0–100% and 100–0% for different w_p, and $w_{SH} = w_S$.

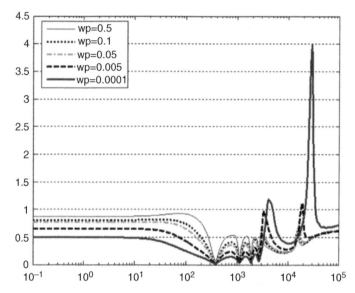

FIGURE 7.11 Upper bound of the structured singular values for different w_p, and $w_{SH} = w_s$.

cases with acceptable transient performances—do not achieve robust stability, with the peak values of $\mu_\Delta[M\,(j\omega)]$ being 1.1 and 4, respectively.

Consider now the following cases of scalars weighting:

Case 6: $w_p = 0.5$, $w_s = 5 \times 10^5$, and $w_{SH} = 0.01 \times w_s$
Case 7: $w_p = 0.1$, $w_s = 5 \times 10^5$, and $w_{SH} = 0.01 \times w_s$
Case 8: $w_p = 0.05$, $w_s = 5 \times 10^5$, and $w_{SH} = 0.01 \times w_s$
Case 9: $w_p = 0.005$, $w_s = 5 \times 10^5$, and $w_{SH} = 0.01 \times w_s$
Case 10: $w_p = 0.0001$, $w_s = 5 \times 10^5$, and $w_{SH} = 0.01 \times w_s$

In cases 6–10 the w_P is assigned the same value as in Cases 1–5, respectively, but smaller weighting is used for the harmonic compensator states as compared to the fundamental's. Figure 7.12 shows the RMS variations for all cases. Comparing Fig. 7.10 and Fig. 7.12, it can be seen that the transient response improves for each of the last five cases, with Case 9 and Case 10 now result in only 2% and 1% RMS variations respectively.

Figure 7.13 shows the upper bound of the structured singular value in Cases 6–10. It can be seen that reducing w_{SH} improves the stability robustness of each of the cases previously considered. The counterpart of Case 4 (Case 9) now achieves robust stability with $\mu_\Delta[M\,(j\omega)]$ peak value of 0.7 giving robust stability margin of $1/0.7 > 1$. Case 10 still does not achieve robust stability with peak value of $\mu_\Delta[M\,(j\omega)]$ still greater than one.

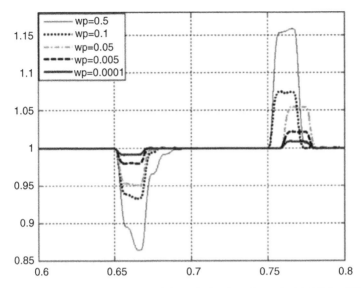

FIGURE 7.12   RMS output voltage variations during 0–100% and 100–0% for different w_p, and $w_{SH} = 0.01 \times w_S$.

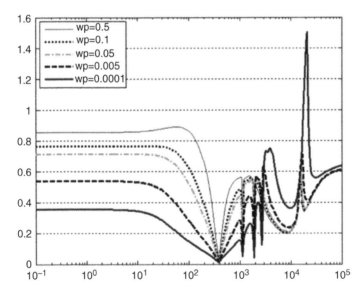

FIGURE 7.13   Upper bound of the structured singular values for different w_p, and $w_{SH} = 0.01 \times w_S$.

Comparing all the above cases, it can be seen that Case 9 gives the best transient performance while still maintaining stability robustness of the system under the structured perturbations considered.

Continuing with the analysis, it is instructive to see how each parameter perturbation affects the stability robustness of the system. For this purpose, the frequency response of each individual perturbation can be plotted as shown in Fig. 7.13 for the system with controller in Case 9. It can be seen from Fig. 7.14 that for all the pertur-

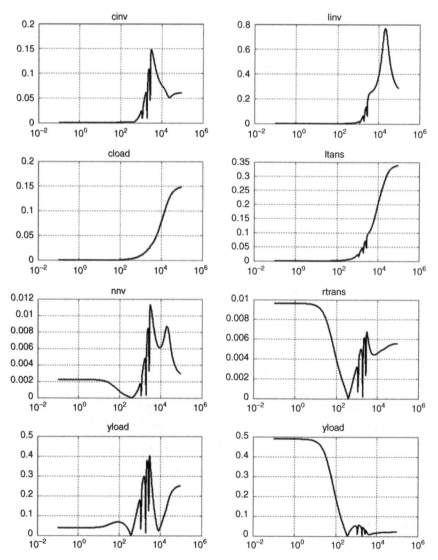

FIGURE 7.14 Individual-perturbation frequency response for system with controller in Case 9. (x-axis in each plot is frequency in rad/sec.)

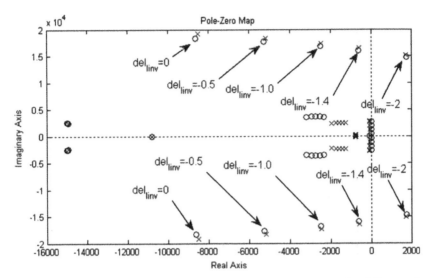

FIGURE 7.15 Pole (**x**)–zero (o) map of closed loop uncertain system for different values of individual-perturbation in inverter filter inductance parameter.

bations, the frequency response values are less than 1.0, which confirms the stability robustness result of the structured singular value presented earlier. The inverter filter inductor has the highest peak frequency response value, with peak value of around 0.7, which means that an individual perturbation in this parameter will be the closest to the margin of making the system to be unstable. Closed-loop pole-zero map of the uncertain closed-loop system in Fig. 7.8 with all other δ's set to zero except for the δ_{linv} can be obtained for different values of $|\delta_{linv}|$ less than 1 and greater than 1. This is illustrated in Fig. 7.15. The system is stable at the nominal inductance value ($\delta_{linv} = 0$), $\delta_{linv} = -0.5$, and at the lowest value of the component tolerance ($\delta_{linv} = -1.0$). As the value δ_{linv} is decreased further, the system finally becomes unstable at δ_{linv} slightly less than -1.4.

The stability robustness of the control system proposed in chapter was verified using the structured Singular Value Method under structured perturbations due to component parameter errors and linear load variations. It was shown that the scalar weighting in the optimal control cost function provides a way of tuning the transient performance of the controller while maintaining stability robustness of the system under perturbations due to plant parameter uncertainties.

CHAPTER 8

PWM RECTIFIER CONTROL FOR THREE-PHASE DISTRIBUTED GENERATION SYSTEM

PWM rectifiers are widely used in three-phase AC–DC–AC systems due to their capability in DC voltage boost and regulation, input power factor correction, and input current harmonic control. However, with the conventional rectifier control technique, the input current tends to be unbalanced under unbalanced inverter load, which contaminates input power source and is therefore undesirable. In this chapter, the cause of the unbalancing is disclosed by evaluating the spectra of the switching functions of the full-bridge three-phase inverter analytically using Bessel function under standard space vector PWM switching scheme, which relates the DC link current and voltage ripples to the inverter load balancing. The analysis shows that the DC link voltage contains a significant second-order harmonic component that affects the voltage loop of the rectifier controller, especially when the control gain is high. A notch filter-based voltage control loop is proposed to eliminate the second harmonic component in the DC-link voltage feedback signal and achieve balanced three-phase input currents. Simulation and experimental results are presented to demonstrate the effectiveness of the proposed control technique in decoupling the rectifier and the inverter under unbalanced load.

8.1 INTRODUCTION

Three-phase AC–DC–AC voltage stiff systems consisting of a front-end rectifier, a dc link with a capacitor, and an inverter are widely used in motor drive, on-line

Integration of Green and Renewable Energy in Electric Power Systems. By A. Keyhani, M. N. Marwali, and M. Dai
Copyright © 2010 John Wiley & Sons, Inc.

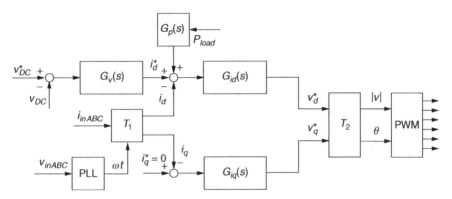

FIGURE 8.1   A conventional PWM rectifier control system with load power feedforward.

UPS, and distributed generation systems. A PWM-controlled front-end rectifier is desirable due to its capability of DC voltage boost and regulation, input power factor correction (PFC), and input current harmonic control. Conventional PFC boost rectifier controllers usually have two feedback loops: an outer voltage loop and an inner current loop, where the voltage regulator generates current command for d-axis current while the q-axis current command is zero for unity power factor control as shown in Fig. 8.1. Under normal operating conditions, steady-state DC bus voltage is a constant and the voltage regulator output (i.e., the d-axis current command) is also a constant, which yields a constant power drawn from the input AC stage and balanced three-phase input currents. However, once the inverter load is not balanced, the output power is no longer a constant, which leads to fluctuation of the DC link voltage. On the rectifier side, the ripple corrupted DC link voltage is fed back to the voltage regulator which generates a fluctuating d-axis current command under a constant DC voltage reference. If the current regulator of d-axis has high bandwidth, it yields fast current tracking and consequently a fluctuating rectifier output current which causes unbalanced front-end input current and high total harmonic distortion (THD) in the input current. This is considered contamination of power system if the front end is fed by utility and therefore undesirable.

The lower the DC coupling capacitance, the better the instantaneous power balance the system could yield. However, this is only desirable under balanced load. Once the inverter load is unbalanced, it is apparent that the steady-state inverter output power is no longer a constant, and neither is the inverter input DC power.

Due to the existence of the DC link capacitor, the rectifier side steady-state power can be decoupled from the inverter side and controlled to be a constant without being affected by the power fluctuation on the inverter side, which will lead to balanced front-end three-phase input currents. To achieve this goal, a new rectifier controller has to be designed to enforce constant input power and not to respond to the DC voltage ripple.

In this research, a switching function concept will be used under standard space vector PWM to quantify the harmonic components in the dc link. According to the analysis result, a notch filter will be designed and applied to eliminate the undesired harmonic component from the feedback signal. The proposed controller with the notch filter will yield constant rectifier power under steady state and balanced three-phase input currents.

Both simulation and experimental test results will be presented to demonstrate the effectiveness of the proposed control technique.

8.2 SYSTEM ANALYSIS

In Fig. 8.2 a standard on-line UPS OR DG system is analyzed. This system consists of a boost type front-end rectifier, a dc link, and a voltage source inverter. Both power converters use standard space vector PWM switching scheme. The output power feedforward technique is not included as a part of the solution of the control problem raised in this problem due to the following two considerations. One is that independent control of the rectifier and inverter leads to lower cost and higher reliability, whereas the other, which is more critical, is that the objective of the power feedforward technique is to achieve fast DC bus voltage regulation, which is inherently against the control goal here. The control goal in this problem is to eliminate the effects of unbalanced three-phase inverter load on the front-end input current and still guarantee unity power factor and fast DC bus voltage regulation against load disturbances.

The effect of load balancing is analyzed as follows. The inverter input current is

$$i_{inv} = S_A i_{outA} + S_B i_{outB} + S_C i_{outC}, \qquad (8.1)$$

where SA, SB, and SC are the switching functions of the top switches of the three inverter legs. When the top switch of leg i is on, $Si = 1$; otherwise, $Si = 0$, where $i \in \{A, B, C\}$. Expand the spectra of these switching functions assuming purely

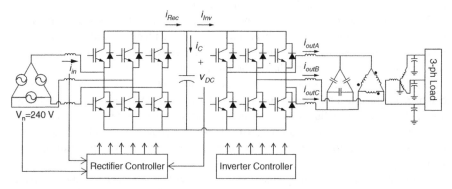

FIGURE 8.2 The three-phase AC–DC–AC system topology.

sinusoidal phase currents as

$$i_{inv}(t) = \sum_{k=1}^{\infty} A_k \sin k\omega t \cdot I_{outA} \sin(\omega t + \phi_A)$$

$$+ \sum_{k=1}^{\infty} A_k \sin k(\omega t - 120°) \cdot I_{outB} \sin(\omega t - 120° + \phi_B)$$

$$+ \sum_{k=1}^{\infty} A_k \sin k(\omega t - 120°) \cdot I_{outC} \sin(\omega t - 120° + \phi_C), \qquad (8.2)$$

where A_k is the magnitude of the kth-order component and $A_k \equiv 0$ for all even k. After trigonometric transform, it can be obtained that

$$i_{inv}(t) = \frac{I_{outA}}{2} \sum_{k=1}^{\infty} A_k\{\cos[(k-1)\omega t - \phi_A] - \cos[(k+1)\omega t + \phi_A]\}$$

$$+ \frac{I_{outB}}{2} \sum_{k=1}^{\infty} A_k\{\cos[(k-1)\omega t - (k-1)120° - \phi_B]$$

$$- \cos[(k+1)\omega t - (k-1)120° - \phi_B]\}$$

$$+ \frac{I_{outC}}{2} \sum_{k=1}^{\infty} A_k\{\cos[(k-1)\omega t + (k-1)120° - \phi_C]$$

$$- \cos[(k+1)\omega t + (k+1)120° + \phi_C]\}$$

$$= I_{inv0} + \sum_{n=1}^{\infty} I_{invn} \sin n(\omega t + \theta_n), \qquad (8.3)$$

where I_{inv0} is the DC component of the inverter input current and I_{invn} is the magnitude of nth-order component of the current. From Eq. (8.3), it is apparent that $I_{outA} = I_{outB} = I_{outC}$ and $\phi_A = \phi_B = \phi_C$ hold and $I_{invn} = 0$ for $n > 0$ if the three-phase load currents are balanced. Otherwise, AC components exist in the current and cause ripple.

It can be observed from Eq. (8.3) that, given fixed three-phase currents, I_{inv0} is proportional to A_1 only, I_{inv2} is a linear combination of A_1 and A_3, I_{inv4} is a linear combination of A_3 and A_5, and so on. Within the low frequency range, $I_{invn} = 0$ for odd n since $A_k \equiv 0$ for even k.

The values of A_k must be calculated to evaluate the significance of each order of the harmonics. Under the standard space vector PWM, the first few components can be calculated using the equation (8.9).

$$A_{2q+1} = \frac{4}{\pi(2q+1)\frac{\omega_m}{\omega_e}} \times J_{2q+1}\left(\frac{\pi a}{2}(2q+1)\frac{\omega_m}{\omega_c}\right), \qquad (8.4)$$

where $q = 0, 1, 2, \ldots, \infty$, ωm is the modulation frequency, ωc is the carrier frequency, $\omega m \ll \omega c$, a is the modulation index, and $Jv(z)$ is the first kind of Bessel function. Equation (8.4) is valid only at the frequency range much less than the carrier frequency where the side band of the carrier frequency is negligible. In the system studied in this research, $\frac{\omega_m}{\omega_c} = \frac{1}{90}$ is applied and the first several terms are calculated assuming normalized modulation index: $A_1 = 1$, $A_3 = 1.142 \times 10^{-4}$, $A_5 = 3.020 \times 10^{-8}$, and $A_7 = 1.030 \times 10^{-11}$. Apparently, compared to A_1, all other terms can be ignored. Based on the above analysis, it can be seen that the A_1 terms only contributes to I_{inv0} and I_{inv2}, where the latter is not desired. Therefore, an undesirable second harmonic component must be generated in i_{inv} by the load unbalance, and the magnitude of the harmonic is determined by the significance of the unbalance.

The second harmonic current fed into the inverter causes fluctuation of the capacitor voltage in the same order. This fluctuation is fed back to the rectifier side voltage regulator that generates a fluctuating d-axis current command $i*d$ as shown in Fig. 8.1. Usually the current regulator has high bandwidth and yields fast response which causes a second harmonic component in the rectifier output current i_{rec}. Due to the same theory as analyzed above, this second harmonic will cause unbalance in the front-end three-phase input currents.

Even though the second harmonic ripple on the dc link voltage seems annoying, suppressing it would not solve the front-end current unbalance problem since the current i_{rec} is still distorted. The strategy proposed in this chapter is to leave the voltage ripple uncontrolled but remove the second harmonic from i_{rec}. However, the rectifier voltage regulation still has to respond to disturbances on the DC bus voltage other than the second harmonic caused by the unbalanced load. Therefore, the second harmonic component needs to be identified and removed from the feedback.

8.3   THE CONTROL STRATEGY

In power supply applications, the nominal frequency is 60 Hz. With a 60-Hz fundamental inverter voltage output, the DC bus ripple must be 120 Hz. A digital band-stop filter can be designed to suppress the ripple with known frequency. A digital Butterworth filter with order $2n$ and lower and upper cutoff frequencies ω_1 and ω_2 can be designed using MATLAB$^\circledR$ function butter(). This filter will be applied to the measured v_{dc}, and its output will be used by the voltage regulator.

A PI controller is used for voltage regulation, and a discrete-time sliding mode controller (DSMC) is used for the inner current loop because it has been proved more effective than PI. The DSMC is described as follows.

The rectifier circuit including the input inductance shown in Fig. 8.2 can be modeled as an LTI system and represented in state space. The current controller controls the input current—that is, the inductor current. In discrete time, the system can be described by

$$i_{in}(k + 1) = A_i i_{in}(k) + B_i v_{pwm}(k) + E_i v_{in}(k), \tag{8.5}$$

where the input current i_{in}, rectifier control voltage v_{pwm}, and input supply voltage v_{in} are all represented in synchronous dq reference frame and A_i, B_i, and E_i are all system coefficients determined by the circuit parameters. Given inverter current command $i_{ref}(k)$, the DSMC equivalent control law is

$$v_{pwm,eq}(k) = B_i^{-1}[i_{ref}(k) - A_i i_{in}(k) - E_i v_{in}(k)]. \tag{8.6}$$

With the rectifier control voltage limit v_{max} determined by the DC bus voltage and the PWM technique used, the actual control voltage becomes

$$v_{inv} = \begin{cases} v_{pwm,eq} & \text{if} \quad ||v_{pwm,eq}|| \leq v_{max}, \\ \dfrac{v_{max}}{||v_{pwm,eq}||} v_{pwm,eq} & \text{otherwise.} \end{cases} \tag{8.7}$$

8.4 SIMULATION RESULTS

Simulations have been performed to compare the performance of the proposed control technique to that of the conventional control. An unbalanced load, where $PA = 1.6$ kW (i.e., approximately 100% load on one phase) and $PB = PC = 0$ is applied to a 5-kVA system as shown in Fig. 8.2. The notch filter design parameters are $n = 2$, $\omega_1 = 140\pi$, and $\omega_2 = 540\pi$. Some steady-state performances are recorded in Table 8.1, where the rectifier three-phase input currents are measured and converted to positive–negative–zero sequence frame to observe the significance of unbalance, where I_{in}^+ and I_{in}^- are the positive and negative sequence quantities, respectively. THD of the input current is also monitored. Table 8.1 shows that the undesirable negative sequence component in the input current has been nearly eliminated and that the input current THD is also reduced under certain amount of load. This result implies that the decoupling between the inverter and the rectifier is achieved by the DC link in that load imbalance on the inverter side does not affect the front-end rectifier and its input current.

TABLE 8.1 Steady-State Simulations, I_{in}^+ and I_{in}^-, Are the Rectifier Input Positive and Negative Sequence Currents, and $I_{in}A$ THD Is the Total Harmonic Distortion of Phase A Input Current

	Conventional	Proposed
I_{in}^+ (A)	7.602	7.566
I_{in}^- (A)	1.486	0.085
I_{inA} THD	5.92%	4.48%

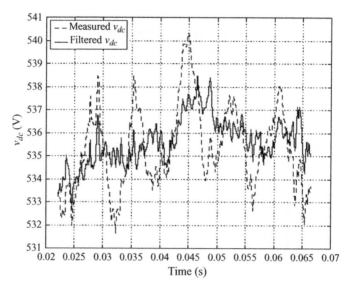

FIGURE 8.3   Experimental results: Measured versus filtered v_{dc}.

The dynamic performance of the proposed control with harmonic compensation is also compared to that of the convention controller as shown in Figs. 8.3 and 8.4, where phase A load steps up from 0%–100%.

By comparing the v_{DC} traces in these figures, it can be observed that the proposed control technique yields balanced three-phase input currents with low THD and simultaneously perform fast v_{DC} regulation under significant load disturbances.

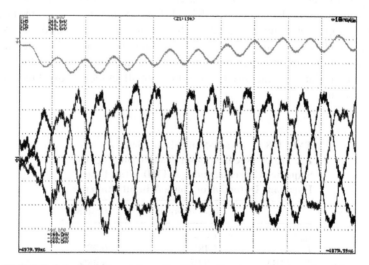

FIGURE 8.4   Experimental results: Conventional, v_{dc} (**top trace**) and i_{inABC}.

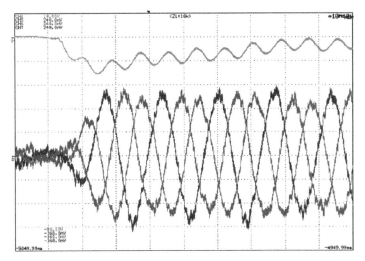

FIGURE 8.5   Experimental results: Proposed, v_{DC} (**top trace**) and i_{inABC}.

8.5 EXPERIMENTAL RESULTS

Experimental tests have been performed on a 5-kVA laboratory setup that has the topology in Fig. 8.2. Two Texas Instruments® TMS320LF2407A DSPs are used to control the rectifier and inverter, respectively.

The same load as used above in the simulation has also been applied in the experiments. Figure 8.3 shows the data of measured v_{dc} and the filtered values collected online under steady state. It is apparent that the directly measured v_{dc} represented by the dashed trace has a dominant 120-Hz component which is significantly suppressed by the fourth-order notch filter as illustrated by the solid trace with existence of measurement noise. Transient tests as done above in the simulation have also been duplicated on the experimental setup and the results are presented by Figure 8.4 and Figure 8.5. It can be observed that the proposed control technique improves the balance of the three-phase input currents while maintaining the unaffected v_{dc} regulation capability with regard to transient load disturbances. The resemblance of the waveforms to those in Figs. 8.6 and 8.7 has validated the simulation results.

8.6 SUMMARY

A novel control technique has been proposed for the front-end PWM rectifier in standard rectifier-inverter systems to achieve decoupling between the converters with the capacitive DC link under unbalanced three-phase inverter load. Analytical work has been conducted to disclose the mechanism regarding how the inverter load unbalancing affects the control of the front-end rectifier. On basis of the analysis, a

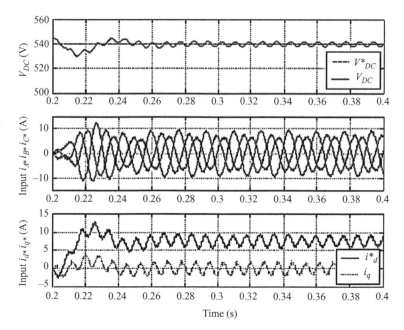

FIGURE 8.6   Simulation results: Transients under conventional control.

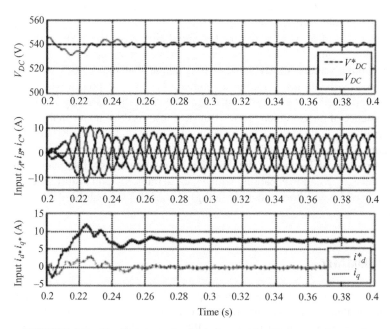

FIGURE 8.7   Simulation results: Transients under the proposed control.

notch filter is then designed and utilized in the voltage loop to suppress the undesired second harmonic component in the DC link voltage feedback, which implements the power decoupling between the rectifier and the inverter and yields balanced input current and does not undermine the dynamic response of DC bus voltage regulation. The effectiveness of the proposed control technique has been demonstrated by both simulations and experimental tests.

CHAPTER 9

MATLAB SIMULINK SIMULATION TESTBED

In this chapter the MATLAB Simulink simulation testbed is presented. The testbed can be used by students to understand the concepts presented in this book. Instructors can use the testbed to design new projects for students as homework problems.

9.1 MATLAB TESTBED FOR CHAPTER 2

```
%%%%%%%%%%%%%%%%%%%%%%%%%%%%%%%%%%%%%%%%%%%%%%%%%%%%%%%%%%%%%%%%%%%%%%%%%%%%
%%   Program to calculate Control Gains using RSP Voltage Controller and   %%
%%   discrete Sliding Mode Current Controller for Impulse                  %%
%%   and to initializes matrices parameters for simulations   %%
%%   Author:  Nanda Marwali                                   %%
%%%%%%%%%%%%%%%%%%%%%%%%%%%%%%%%%%%%%%%%%%%%%%%%%%%%%%%%%%%%%%%%%%%%%%%%%%%%

clear all

% Define fundamental output frequency;
ffun=60;

wfun=2*pi*ffun;


% Define control sampling time
Tsamp=1/60/51;    %6536/20e6;
```

Integration of Green and Renewable Energy in Electric Power Systems. By A. Keyhani, M. N. Marwali, and M. Dai
Copyright © 2010 John Wiley & Sons, Inc.

```
Cinv=540e-6;                      % Inverter capacitor filter
Linv=298e-6;                      % Inverter inductor filter
Cload=90e-6;                      % Grass capacitor
Ltrans=0.03*208*208/80e3/wfun;    % 3% p.u transformer inductance
Rtrans=0; %0.03*208*208/80e3;     % 3% p.u transformer resistance
tr=120/245;                       % Inverter To Output Turn ratio
Ilimit=3*80e3/245*sqrt(2);        % 300% inverter current limit in Line-
Line Amp peak


%% Define harmonics frequencies
w1=wfun;
w2=2*wfun;
w3=3*wfun;
w5=5*wfun;
w4=4*wfun;
w7=7*wfun;
T1=1/ffun;

% Define Delta-Wye Transformer Voltages and Currents Transfer Matrices

Tri_qd=3/2*tr*[1 sqrt(3)  ;
             -sqrt(3) 1 ];

Trv_qd=1/2*tr*[1 -sqrt(3);
        sqrt(3) 1 ];

%%%%%%%%%%%%%%%%%%%%%%%%%%%%%%%%%%%%%%%%%%%%%%%%%%%%%%%%%%%%%%%%%
% Start design of discrete SM current controller
%%%%%%%%%%%%%%%%%%%%%%%%%%%%%%%%%%%%%%%%%%%%%%%%%%%%%%%%%%%%%%%%%

%% Define the plant for the current controller

A=[zeros(2,2)            eye(2,2)/(3*Cinv) ;
   -1/Linv*eye(2)        -0.01/Linv*eye(2)];


B=[zeros(2,2);
   1/Linv*eye(2)];

E=[-Tri_qd/(3*Cinv);
   zeros(2,2)];

F=[zeros(2,2);
   -eye(2)];

C=[zeros(2,2) eye(2)];

D=zeros(2,2);

%% Discretize the plant for the current controller

sysc=ss(A,B,C,zeros(size(C,1),size(B,2)));

sysd=c2d(sysc,Tsamp,'zoh');

[Acurrd,Bcurrd,Ccurrd,Dcurrd]=ssdata(sysd);

CBinv=inv(Ccurrd*Bcurrd);
```

```
CA=Ccurrd*Acurrd;

CD=Ccurrd*F;

sysc=ss(A,E,C,zeros(size(C,1),size(B,2)));

sysd=c2d(sysc,Tsamp,'zoh');

[Acurrd1,Ecurrd,Ccurrd1,Dcurrd1]=ssdata(sysd);

CE=Ccurrd1*Ecurrd;


%%% End Discrete sliding mode controllers gains calculation


%%%%%%%%%%%%%%%%%%%%%%%%%%%%%%%%%%%%%%%%%%%%%%%%%%%%%%%%%%%%%%%%
% Start design of Perfect RSP voltages controllers
%%%%%%%%%%%%%%%%%%%%%%%%%%%%%%%%%%%%%%%%%%%%%%%%%%%%%%%%%%%%%%%%

%% Define the true plant

Ao=[   zeros(2,2)          eye(2,2)/(3*Cinv)   zeros(2,2)      -Tri_qd/(3*Cinv);
      -eye(2,2)/Linv       zeros(2,2)          zeros(2,2)      zeros(2,2);
       zeros(2,2)          zeros(2,2)          zeros(2,2)      eye(2,2)/Cload;
       Trv_qd/Ltrans       zeros(2,2)         -eye(2,2)/Ltrans  -
Rtrans*eye(2,2)/Ltrans];


Bo=[zeros(2,2);
    eye(2,2)/Linv;
    zeros(2,2);
    zeros(2,2)];

Co=[zeros(2,2) zeros(2,2) eye(2) zeros(2,2)];

%% Define the analog servo compensator

Ch0=zeros(2,2);

Ch1=[zeros(2,2)   eye(2);
    -w1^2*eye(2) zeros(2,2)];

Ch5=[zeros(2,2)   eye(2);
    -w5^2*eye(2) zeros(2,2)];

Ch7=[zeros(2,2)   eye(2);
    -w7^2*eye(2) zeros(2,2)];


Ch_star=[Ch1                zeros(size(Ch1))   zeros(size(Ch1)) ;
    zeros(size(Ch1))        Ch5                zeros(size(Ch1)) ;
    zeros(size(Ch1))        zeros(size(Ch1)) Ch7     ];


Bh_star=[zeros(2,2);
         eye(2,2);
         zeros(2,2);
         eye(2,2);
```

```
        zeros(2,2);
        eye(2,2)];


% Discretize true plant
sysc=ss(Ao,Bo,Co,zeros(size(C,1),size(B,2)));

sysd=c2d(sysc,Tsamp,'zoh');

[Aod,Bod,Cd,Dd]=ssdata(sysd);


% Calculate equivalent plant+DSM current controller

C1=[eye(4)  zeros(4,4)];

C2=[zeros(2,4)  eye(2) zeros(2,2)];

Ad=Aod-Bod*(CBinv*CA*C1+CBinv*CE*C2);

Bd=Bod*CBinv;


% Discetize the controller
csysc=ss(Ch_star,Bh_star,eye(size(Ch_star,1)),zeros(size(Ch_star,1),
    size(Bh_star,2)));

[csysbc,T]=ssbal(csysc);

csysd=c2d(csysbc,Tsamp,'zoh');

[Acon_d,Bcon_d,Ccon_d,Dcon_d]=ssdata(csysd);


% Form the augmented equivalent plant and the servocompensator

Ad_big=[Ad     zeros(size(Ad,1),size(Acon_d,2));
        -Bcon_d*Cd         Acon_d];

Bd_big=[Bd ; -Bcon_d*Dd];


% Define the weighting matrices

epsilon=1e-5;
Q2=eye(size(Acon_d,1));

state_W=0.2; % 80 kVA

Q1=state_W*eye(size(Ad,1));

Q=[    Q1                   zeros(size(Ad,1),size(Acon_d,2));
   zeros(size(Acon_d,1),size(Ad,2))            5e5*Q2];


R=epsilon*eye(2);

% Now performed the optimal calculations of the gains
[Kd,S,E]=dlqr(Ad_big,Bd_big,Q,R);
Kd=-Kd;
```

```
%%%%%%%%%%%%%%%%%%%%%%%%%%%%%%%%%%%%%%%%%%%%%%%%%%%%%%%%%%%%%%%%%%%
% Create matrices parameters for the plant states equations
%%%%%%%%%%%%%%%%%%%%%%%%%%%%%%%%%%%%%%%%%%%%%%%%%%%%%%%%%%%%%%%%%%%

Rfl=208*208/(80e3*0.8);
Rtrans=0.03*208*208/80e3;

Tri=[1 -2 1;
     1  1 -2;
    -2  1 1];

Trv=[0 0 -1;
    -1 0 0;
     0 -1 0];


Tri_qd0=3/2*tr*[1 sqrt(3) 0;
             -sqrt(3) 1 0];

Trv_qd0=1/2*tr*[1 -sqrt(3);
        sqrt(3) 1 ;
      0    0 ];

Ks=2/3*[cos(0) cos(0-2*pi/3) cos(0+2*pi/3);
        sin(0) sin(0-2*pi/3) sin(0+2*pi/3);
           1          1           1    ];

Ksinv=inv(Ks)  ;


% Define the B and D matrices

Bsim=[zeros(2,2);
     eye(2,2)/Linv;
     zeros(3,2);
     zeros(3,2)];

Dsim=zeros(13,2);


% Full load

Rload33=[ Rfl   0   0;
          0    Rfl  0;
          0    0   Rfl];

Afl=[  zeros(2,2)          eye(2,2)/(3*Cinv) zeros(2,3)      -Tri_qd0/(3*Cinv);
     -eye(2,2)/Linv        zeros(2,2)        zeros(2,3)    zeros(2,3);
      zeros(3,2)           zeros(3,2)       -Ks*inv(Rload33)*Ksinv/Cload
eye(3,3)/Cload;  Trv_qd0/Ltrans      zeros(3,2)
-eye(3,3)/Ltrans  -Rtrans*eye(3,3)/Ltrans];


Cfl=[ [eye(10)];
      [zeros(3,4) Ks*inv(Rload33)*Ksinv zeros(3,3)]];


% No load
```

```
Rload33=[ 100    0    0;
            0   100    0;
            0    0   100];

An1=[  zeros(2,2)         eye(2,2)/(3*Cinv)  zeros(2,3)      -Tri_qd0/(3*Cinv);
       -eye(2,2)/Linv     zeros(2,2)         zeros(2,3)      zeros(2,3);
        zeros(3,2)        zeros(3,2)         -Ks*inv(Rload33)*Ksinv/Cload
eye(3,3)/Cload;       Trv_qd0/Ltrans     zeros(3,2)
-eye(3,3)/Ltrans  -Rtrans*eye(3,3)/Ltrans];

Cn1=[ [eye(10)];
      [zeros(3,4) Ks*inv(Rload33)*Ksinv zeros(3,3)]];


% Single phase loaded

Rload33=[ Rfl    0    0;
            0   100    0;
            0    0   100];

Ap1=[  zeros(2,2)         eye(2,2)/(3*Cinv)  zeros(2,3)      -Tri_qd0/(3*Cinv);
       -eye(2,2)/Linv     zeros(2,2)         zeros(2,3)      zeros(2,3);
        zeros(3,2)        zeros(3,2)         -Ks*inv(Rload33)*Ksinv/Cload
eye(3,3)/Cload;       Trv_qd0/Ltrans     zeros(3,2)
-eye(3,3)/Ltrans  -Rtrans*eye(3,3)/Ltrans];

Cp1=[ [eye(10)];
      [zeros(3,4) Ks*inv(Rload33)*Ksinv zeros(3,3)]];


% Two phase loaded

Rload33=[ Rfl    0    0;
            0   Rfl    0;
            0    0   100];

Ap2=[  zeros(2,2)         eye(2,2)/(3*Cinv)  zeros(2,3)      -Tri_qd0/(3*Cinv);
       -eye(2,2)/Linv     zeros(2,2)         zeros(2,3)      zeros(2,3);
        zeros(3,2)        zeros(3,2)         -Ks*inv(Rload33)*Ksinv/Cload
eye(3,3)/Cload;       Trv_qd0/Ltrans     zeros(3,2)
-eye(3,3)/Ltrans  -Rtrans*eye(3,3)/Ltrans];

Cp2=[ [eye(10)];
      [zeros(3,4) Ks*inv(Rload33)*Ksinv zeros(3,3)]];


% 500% load
Rload33=[ sqrt(0.2*Rfl)    0    0;
            0   sqrt(0.2*Rfl)    0;
            0    0   sqrt(0.2*Rfl)];

Aov=[  zeros(2,2)         eye(2,2)/(3*Cinv)  zeros(2,3)      -Tri_qd0/(3*Cinv);
       -eye(2,2)/Linv     zeros(2,2)         zeros(2,3)      zeros(2,3);
        zeros(3,2)        zeros(3,2)         -Ks*inv(Rload33)*Ksinv/Cload
eye(3,3)/Cload;       Trv_qd0/Ltrans     zeros(3,2)
-eye(3,3)/Ltrans  -Rtrans*eye(3,3)/Ltrans];

Cov=[ [eye(10)];
      [zeros(3,4) Ks*inv(Rload33)*Ksinv zeros(3,3)]];
```

```
% Short circuit

Rload33=[ 0.001   0    0;
            0    0.001   0;
            0     0    0.001];

Asc=[   zeros(2,2)          eye(2,2)/(3*Cinv)   zeros(2,3)      -Tri_qd0/(3*Cinv);
       -eye(2,2)/Linv        zeros(2,2)         zeros(2,3)      zeros(2,3);
         zeros(3,2)          zeros(3,2)        -Ks*inv(Rload33)*Ksinv/Cload
eye(3,3)/Cload;        Trv_qd0/Ltrans        zeros(3,2)
-eye(3,3)/Ltrans   -Rtrans*eye(3,3)/Ltrans];

Csc=[ [eye(10)];
       [zeros(3,4) Ks*inv(Rload33)*Ksinv zeros(3,3)]];
```

9.2 FILE SFUNFFT.M

```
function  [sys, x0, str, ts] = sfunfft(t,x,u,flag,fftpts,npts,HowOften,...
                                offset,ts,averaging,nharmonics)
%SFUNPSD an S-function which performs spectral analysis using ffts.
%     This M-file is designed to be used in a Simulink S-function block.
%     It stores up a buffer of input and output points of the system
%     then plots the power spectral density of the input signal.
%
%     The input arguments are:
%     npts:          number of points to use in the fft (e.g. 128)
%     HowOften:      how often to plot the ffts (e.g. 64)
%     offset:        sample time offset (usually zeros)
%     ts:            how often to sample points (secs)
%     averaging:     whether to average the psd or not
%
%     Two or three plots are given: the time history, the instantaneous psd
%     the average psd.
%
%     See also, FFT, SPECTRUM, SFUNTMPL, SFUNTF.

%     Copyright (c) 1990-97 by The MathWorks, Inc.
%     $Revision: 1.25 $
%     Andrew Grace 5-30-91.
%     Revised Wes Wang 4-28-93, 8-17-93.
%     Revised Charlie Ko 9-26-96.


switch flag

  %%%%%%%%%%%%%%%%%%%%
  % Initialization %
  %%%%%%%%%%%%%%%%%%%%%
  case 0
    [sys,x0,ts] = mdlInitializeSizes(fftpts,npts,HowOften,offset,...
                                ts,averaging);
    SetBlockCallbacks(gcbh);
```

```
%%%%%%%%%%
% Update %
%%%%%%%%%%
case 2
   sys = mdlUpdate(t,x,u,fftpts,npts,HowOften,offset,ts,averaging,
         nharmonics);
case 3
   nstates = npts + 2 + averaging * round(fftpts/2) + 1;
   sys=x(nstates+1);
%%%%%%%%%%%%%%%%%%%%%%%%%
% GetTimeOfNextVarHit %
%%%%%%%%%%%%%%%%%%%%%%%%%
case 4
   sys=mdlGetTimeOfNextVarHit(t,x,u,ts);

%%%%%%%%%
% Start %
%%%%%%%%%
case 'Start'
   LocalBlockStartFcn

%%%%%%%%%%%%%%
% NameChange %
%%%%%%%%%%%%%%
case 'NameChange'
   LocalBlockNameChangeFcn

%%%%%%%%%%%%%%%%%%%%%%%%%%
% CopyBlock, LoadBlock %
%%%%%%%%%%%%%%%%%%%%%%%%%%
case { 'CopyBlock', 'LoadBlock' }
   LocalBlockLoadCopyFcn

%%%%%%%%%%%%%%%
% DeleteBlock %
%%%%%%%%%%%%%%%
case 'DeleteBlock'
   LocalBlockDeleteFcn

%%%%%%%%%%%%%%%%
% DeleteFigure %
%%%%%%%%%%%%%%%%
case 'DeleteFigure'
   LocalFigureDeleteFcn

%%%%%%%%%%%%%%%%
% Unused flags %
%%%%%%%%%%%%%%%%
case { 3, 9 }
   sys = []; %do nothing

%%%%%%%%%%%%%%%%%%%%
% Unexpected flags %
%%%%%%%%%%%%%%%%%%%%
otherwise
   if ischar(flag),
      errmsg=sprintf('Unhandled flag: ''%s''', flag);
```

```
    else
      errmsg=sprintf('Unhandled flag: %d', flag);
    end

    error(errmsg);

  end

% end sfunpsd

%
%===========================================================================
% mdlInitializeSizes
% Return the sizes, initial conditions, and sample times for the S-function.
%===========================================================================
%
function [sys,x0,ts] = mdlInitializeSizes(fftpts,npts,HowOften,offset,...
                                    ts,averaging)

nstates = npts + 2 + averaging * round(fftpts/2) + 1; % No. of dis-
crete states
initial_states = [1; zeros(nstates-1,1)];          % Initial conditions
if HowOften > npts
  error('The number of points in the buffer must exceed the plot frequency')
end
initial_states(nstates+1) = 0;

sizes = simsizes;
sizes.NumContStates  = 0;
sizes.NumDiscStates  = nstates+1;
sizes.NumOutputs     = 1;
sizes.NumInputs      = 2;
sizes.DirFeedthrough = 0;
sizes.NumSampleTimes = 1;

sys = simsizes(sizes);

x0  = initial_states;
str = [];
%ts  = [ts offset];
ts=[-2 0];
% end mdlInitializeSizes

%
%===========================================================================
% mdlGetTimeOfNextVarHit
% Return the time of the next hit for this block.  Note that the result is
% absolute time.
%===========================================================================
%
function sys=mdlGetTimeOfNextVarHit(t,x,u,ts)
  if u(2)==0
      sys = t + 1/60;
    else
      sys=t+ts;
    end
% end mdlGetTimeOfNextVarHit

%
```

```
%==============================================================================
% mdlUpdates
% Compute Discrete State updates.
%==============================================================================
%
function [sys]=mdlUpdate(t,x,u,fftpts,npts,HowOften,offset,ts,averaging,
                 nharmonics)

nstates = npts + 2 + averaging * round(fftpts/2) + 1; % No. of dis-
crete states

if u(2)==0
   sys=x;
   return
end

%
% Find the figure handle associated with this block. If the handle
% is not valid (i.e. - the user closed the figure during simulation),
% mdlUpdates performs no operations.
%
FigHandle = GetSfunPSDFigure(gcbh);
if ~ishandle(FigHandle)
  sys=[];
  return
end

%
% Initialize the return argument.
%
sys = x;

%
% Increment the counter and store the current input in the
% discrete state referenced by this counter.
%
x(1) = x(1) + 1;
sys(x(1)) = u(1);

if rem(x(1),HowOften)==0
  ud = get(FigHandle,'UserData');
  buffer = [sys(x(1)+1:npts+1);sys(2:x(1))];
  n = fftpts/2;
  freq =(1/ts); % Multiply by 2*pi to get radians
  w = freq*(0:n-1)./fftpts;

  % Detrend the data: remove best straight line fit
  a = [(1:npts)'/npts ones(npts,1)];
  y = buffer;%  - a*(a\buffer);

  % Hammining window to remove transient effects at the
  % beginning and end of the time sequence.
  nw = min(fftpts, npts);
  win = .5*(1 - cos(2*pi*(1:nw)'/(nw+1)));

  g = fft(y(1:nw));
%  g(1)=0.0;
  % Perform averaging with overlap if number of fftpts
  % is less than buffer
```

```
%  ng = fftpts;
%  while (ng >= (npts - fftpts))
%     g = (g + fft(win.*y(ng+1:ng+fftpts),fftpts)) / 2;
%     ng = ng + n; % For no overlap set: ng = ng + fftpts;
%  end

  g = g(1:n);
  %psd = (abs(g).^2)./norm(win)^2;
  apsd=2*abs(g)./fftpts;
  psd=apsd./apsd(2);
  yact=apsd(1:nharmonics);
  ynorm=psd(1:nharmonics);
%  ynorm*100
  THD=sqrt(sum(yact(3:nharmonics).^2)/sum(yact(2:nharmonics).^2))*100;
  sys(nstates+1)=THD;
%  RMS=sqrt(sum(yact(2:nharmonics).^2));
%  test=abs(psd(2))
  tvec = (t - ts * npts + ts * (1:npts));
  % For Time History plot.
  set(ud.TimeHistoryAxis,...
      'Visible','on',...
      'XGrid','on',...
      'YGrid','on',...
      'Xlim',[min(tvec) max(tvec)],...
      'Ylim',[min(buffer*.99) max(buffer*1.01+eps)]);
  set(ud.TimeHistoryInputLine,...
      'XData',tvec,...
      'YData',buffer);
  set(ud.TimeHistoryTitle,...
      'String','Time history');
  xl = get(ud.TimeHistoryAxis,'Xlabel');
  set(xl,...
      'String','Time (secs)');

  if averaging
    cnt = sys(npts+2+n); % Counter for averaging
    sys(npts+2:npts+1+n) = cnt/(cnt+1)*sys(npts+2:npts+1+n)+psd/(cnt+1);
    sys(npts+2+n) = sys(npts+2+n) + 1;
    psd = sys(npts+3:npts+1+n);
    tmp = 'Average Power Spectral Density';
  else
    tmp = 'Spectral Density';
    psd = psd(1:n);
  end
  ysc = psd(~isnan(psd));

  if isempty(ysc)
    ysc=[0 1];
  else
    ysc = sort([min(ysc*.99), max(ysc*1.01+eps)]);
  end;
  % Set up data using fancing indexing
  [mm,nn] = size(w.');
  xx = zeros(3*mm,nn);
  xx(1:3:3*mm,:) = w.';
  xx(2:3:3*mm,:) = w.';
  xx(3:3:3*mm,:) = NaN;

  [mm,nn] = size(psd);
```

```
  yy = zeros(3*mm,nn);
  yy(2:3:3*mm,:) = psd;
  yy(3:3:3*mm,:) = NaN;

  % For the PSD plot.
  set(ud.PSDAxis,...
     'Visible','on',...
     'GridLineStyle',':',...
     'XGrid','on',...
     'YGrid','on',...
     'Xlim',[min(w(1:nharmonics)), max(w(1:nharmonics))],...
     'Ylim',ysc);
  set(ud.PSDInputLine,...
      'XData',xx,...
      'YData',yy,...
      'LineWidth',1.5);
  set(ud.PSDTitle,...
     'String', tmp);
  xl = get(ud.PSDAxis, 'Xlabel');
  set(xl,...
     'String','Frequency (Hz)')

%   phase = (180/pi)*unwrap(atan2(imag(g),real(g)));
%   phase = phase(2:n);
%   ysc = phase(~isnan(phase));
%   if isempty(ysc)
%     ysc=[0 1];
%   else
%     ysc = sort([min(ysc*.99), max(ysc*1.01+eps)]);
%   end;
%   %For phase plot
%   set(ud.PhaseAxis,...
%        'Visible','on',...
%        'Xlim',[min(w(2:n)) max(w(2:n))],...
%        'Ylim',ysc);
%   set(ud.PhaseInputLine,'XData',w(2:n),'YData',phase)
%   set(ud.PhaseTitle, 'String',[tmp '(phase)'])
%   xl = get(ud.PhaseAxis, 'Xlabel');
%   set(xl, 'String', 'Frequency (rads/sec)')
%   yl = get(ud.PhaseAxis, 'Ylabel');
%   set(yl, 'String','Degrees')

drawnow
end

%
% If the buffer is full, reset the counter. The counter is store in
% the first discrete state.
%
if sys(1,1) == npts
  x(1,1) = 1;
end
sys(1,1) = x(1,1);

% end mdlUpdate
```

```
%
%===============================================================================
% LocalBlockStartFcn
% Function that is called when the simulation starts.  Initialize the
% XY Graph scope figure.
%===============================================================================
%
function LocalBlockStartFcn

%
% Retrieve the block's figure handle.
%
FigHandle = GetSfunPSDFigure(gcbh);
if isempty(FigHandle)
  FigHandle=-1;
end

if ~ishandle(FigHandle)
  FigHandle = CreateSfunPSDFigure;
end
ud = get(FigHandle,'UserData');
set(ud.TimeHistoryTitle,'String','Working - please wait');

% end LocalBlockStartFcn

%
%===============================================================================
% CreateSfunPSDFigure
% Creates the figure window that is associated with the PSD Block.
%===============================================================================
%
function FigHandle=CreateSfunPSDFigure

%
% The figure doesn't already exist, create one.
%
%                     'Position',      [100 100 400 510],...

FigHandle = figure('Units',           'points',...
                   'NumberTitle',    'off',...
                   'IntegerHandle','off',...
                   'MenuBar',        'figure',...
                   'Name',            get_param(gcbh,'Name'),...
                   'DeleteFcn',       'sfunfft([],[],[],''DeleteFigure'')');

%
% Store the block's handle in the figure's UserData.
%
ud.Block=gcbh;

%
% Create the various objects within the figure.
%

% Subplot of the the time history data.
ud.TimeHistoryAxis = subplot(211);
%set(ud.TimeHistoryAxis,'Position',[.13 .72 .77 .2]);
ud.TimeHistoryInputLine = plot(0,0,'m','EraseMode','Normal');
```

```
ud.TimeHistoryTitle = get(ud.TimeHistoryAxis,'Title');
set(ud.TimeHistoryAxis,'Visible','off');

% Subplot of the Power Spectral Density.
ud.PSDAxis = subplot(212);
%set(ud.PSDAxis,'Position',[.13 .4 .77 .2]);
ud.PSDInputLine= plot(0,0,'EraseMode','Normal');
ud.PSDTitle = get(ud.PSDAxis,'Title');
set(ud.PSDAxis,'Visible','off');

% Subplot of the phase shift diagram.
%ud.PhaseAxis = subplot(313);
%set(ud.PhaseAxis,'Position',[.13 .08 .77 .2]);
%ud.PhaseInputLine = plot(0,0,'EraseMode','None');
%ud.PhaseTitle = get(ud.PhaseAxis,'Title');
%set(ud.PhaseAxis,'Visible','off');

%
% Place the figure handle in the current block's UserData.  Then
% place the various object handles into the Figure's UserData.
%
SetSfunPSDFigure(gcbh,FigHandle);
set(FigHandle,'HandleVisibility','on','UserData',ud);

% end CreateSfunPSDFigure

%
%=============================================================================
% GetSfunPSDFigure
% Retrieves the figure handle that is associated with this block from the
% block's parent subsystem's UserData.
%=============================================================================
%
function FigHandle=GetSfunPSDFigure(block)

if strcmp(get_param(block,'BlockType'),'S-Function'),
  block=get_param(block,'Parent');
end

FigHandle=get_param(block,'UserData');
if isempty(FigHandle),
  FigHandle=-1;
end

% end GetSfunPSDFigure.

%
%=============================================================================
% SetSfunPSDFigure
% Stores the figure handle that is associated with this block into the
% block's UserData.
%=============================================================================
%
function SetSfunPSDFigure(block,FigHandle)

if strcmp(get_param(block,'BlockType'),'S-Function'),
  block=get_param(block,'Parent');
end
```

```
set_param(block,'UserData',FigHandle);

% end SetSfunPSDFigure.

%
%==============================================================================
% LocalBlockNameChangeFcn
% Function that handles name changes on the PSD Graph scope block.
%==============================================================================
%
function LocalBlockNameChangeFcn

%
% get the figure associated with this block, if it's valid, change
% the name of the figure
%
FigHandle = GetSfunPSDFigure(gcbh);
if ishandle(FigHandle),
  set(FigHandle,'Name',get_param(gcbh,'Name'));
end

% end LocalBlockNameChangeFcn

%
%==============================================================================
% LocalBlockLoadCopyFcn
% This is the PSD block's CopyFcn and LoadFcn.  Initialize the block's
% UserData such that a figure is not associated with the block.
%==============================================================================
%
function LocalBlockLoadCopyFcn

SetSfunPSDFigure(gcbh,-1);

% end LocalBlockLoadCopyFcn

%
%==============================================================================
% LocalBlockDeleteFcn
% This is the PSD block's DeleteFcn.  Delete the block's figure
% window, if present, upon deletion of the block.
%==============================================================================
%
function LocalBlockDeleteFcn

%
% Get the figure handle, the second arg to SfunCorrFigure is set to zero
% so that that function doesn't create the figure if it doesn't exist.
%
FigHandle=GetSfunPSDFigure(gcbh);
if ishandle(FigHandle),
  delete(FigHandle);
  SetSfunPSDFigure(gcbh,-1);
end

% end LocalBlockDeleteFcn

%
%==============================================================================
```

```
% LocalFigureDeleteFcn
% This is the PSD's figure window's DeleteFcn.  The figure window
% is being deleted, update the correlation block's UserData to
% reflect the change.
%=============================================================================
%
function LocalFigureDeleteFcn

%
% Get the block associated with this figure and set it's figure to -1
%
ud=get(gcbf,'UserData');
SetSfunPSDFigure(ud.Block,-1)

% end LocalFigureDeleteFcn

%
%=============================================================================
% SetBlockCallbacks
% This sets the LoadFcn, CopyFcn and DeleteFcn of the block.
%=============================================================================
%
function SetBlockCallbacks(block)

%
% the actual source of the block is the parent subsystem
%
block=get_param(block,'Parent');

%
% if the block isn't linked, issue a warning, and then set the callbacks
% for the block so that it has the proper operation
%
%if strcmp(get_param(block,'LinkStatus'),'none'),
%  warnmsg=sprintf(['The Power Spectral Density Graph scope ''%s'' should ' ...
%                   'be replaced with a new version from the ' ...
%                   'simulink_extras block library'],...
%                   block);
%  warning(warnmsg);

  callbacks={
    'CopyFcn',        'sfunfft([],[],[],''CopyBlock'')' ;
    'DeleteFcn',      'sfunfft([],[],[],''DeleteBlock'')' ;
    'LoadFcn',        'sfunfft([],[],[],''LoadBlock'')' ;
    'StartFcn',       'sfunfft([],[],[],''Start'')' ;
    'NameChangeFcn',  'sfunfft([],[],[],''NameChange'')' ;
  };

  for i=1:length(callbacks),
     if ~strcmp(get_param(block,callbacks{i,1}),callbacks{i,2}),
       set_param(block,callbacks{i,1},callbacks{i,2})
     end
  end
%end

% end SetBlockCallbacks
```

APPENDIX A

SIMULINK MODEL *DSIMSERVO.MDL*

A.1 MODEL: dsimservo

TABLE A.1 Model Functions

Function Name	Parent Blocks	Calling String
eye	dsimservo/Voltage Controller/Servo Compensator	eye(size(Acon_d))
pi	dsimservo/D 5th. harm, dsimservo/D 5th harm, dsimservo/D 7th harm, dsimservo/D 7th harm, dsimservo/D Fund, dsimservo/D Fund, dsimservo/Q 5th harm, dsimservo/Q 7th harm, dsimservo/Q Fundamental	2*pi*ffun*7 -pi/2*7 2*pi*ffun*3 pi/2*3 2*pi*ffun pi/2 2*pi*ffun*7 2*pi*ffun*3 2*pi*ffun
size	dsimservo/Voltage Controller/Servo Compensator	eye(size(Acon_d))
sqrt	dsimservo/Current Controller/PWM command limit/Fcn1, dsimservo/Current Controller/PWM command limit/Fcn2, dsimservo/D 5th harm, dsimservo/D 7th harm, dsimservo/D Fund, dsimservo/Q 5th harm, dsimservo/Q 7th harm, dsimservo/Q Fundamental, dsimservo/Voltage Controller/Current Limit/Fcn1, dsimservo/Voltage Controller/Current Limit/Fcn2	sqrt(u[1]*u[1]+u[2]*u[2]) u[3]/sqrt(u[1]*u[1]+u[2]*u[2]) 245.0*sqrt(2)*0.0 245.0*sqrt(2)*0 120*sqrt(2) 245.0*sqrt(2)*0.0 245.0*sqrt(2)*0 120*sqrt(2) sqrt(u[1]*u[1]+u[2]*u[2]) Ilimit/sqrt(u[1]*u[1]+u[2]*u[2])

Integration of Green and Renewable Energy in Electric Power Systems. By A. Keyhani, M. N. Marwali, and M. Dai
Copyright © 2010 John Wiley & Sons, Inc.

TABLE A.2   Model Variables

Variable Name	Parent Blocks	Calling String
Acon_d	dsimservo/Voltage Controller/Servo Compensator, dsimservo/Voltage Controller/Servo Compensator	Acon_d eye(size(Acon_d))
Afl	dsimservo/Plant	Afl
Bcon_d	dsimservo/Voltage Controller/Servo Compensator	Bcon_d
Bsim	dsimservo/Plant	Bsim
CA	dsimservo/Current Controller/CBinv*CA	CBinv*CA
CBinv	dsimservo/Current Controller/CBinv*CA, dsimservo/Current Controller/CBinv*CE, dsimservo/Current Controller/inv(K)5	CBinv*CA CBinv*CE CBinv
CE	dsimservo/Current Controller/CBinv*CE	CBinv*CE
Cfl	dsimservo/Plant	Cfl
Dcon_d	dsimservo/Voltage Controller/Servo Compensator	Dcon_d
Dsim	dsimservo/Plant	Dsim
Ilimit	dsimservo/Voltage Controller/Constant, dsimservo/Voltage Controller/Current Limit/Constant1, dsimservo/Voltage Controller/Current Limit/Fcn2	Ilimit Ilimit Ilimit/sqrt(u[1]*u[1]+u[2]*u[2])
Kd	dsimservo/Voltage Controller/K	Kd
Ksinv	dsimservo/DQ0Stat2ABC2, dsimservo/DQ0Stat2ABC3, dsimservo/DQ0Stat2ABC4	Ksinv Ksinv 1/3*[1 −1 0;0 1 −1;−1 0 1]*Ksinv(:,1:2)
Tsamp	dsimservo/Current Controller/Unit Delay1, dsimservo/Voltage Controller/Servo Compensator, dsimservo/Voltage Controller/Unit Delay2, dsimservo/Zero-Order Hold5, dsimservo/Zero-Order Hold6, dsimservo/Zero-Order Hold7, dsimservo/Zero-Order Hold8	Tsamp Tsamp Tsamp Tsamp Tsamp Tsamp Tsamp
ffun	dsimservo/RMS, dsimservo/RMS1, dsimservo/RMS2, dsimservo/D 5th harm, dsimservo/D 7th harm., dsimservo/D Fund, dsimservo/Q 5th harm., dsimservo/Q 7th harm, dsimservo/Q Fundamental	ffun ffun ffun 2*pi*ffun*7 2*pi*ffun*3 2*pi*ffun 2*pi*ffun*7 2*pi*ffun*3 2*pi*ffun

TABLE A.3 dsimservo Simulation Parameters

Solver ode15s	*ZeroCross* on	*StartTime* 0.0 *StopTime* 100
RelTol 1e-6	*AbsTol* auto	*Refine* 1
InitialStep auto	*FixedStep* auto	*MaxStep* 10e-6
LimitMaxRows off	*MaxRows* 1000	*Decimation* 1

TABLE A.4 Constant Block Properties

Name	Value
Constant3	540
Constant1	0
Constant2	1
Constant	I_{limit}
Constant1	0
Constant1	I_{limit}

TABLE A.5 Demux Block Properties

Name	Outputs
Demux	3
Demux2	2

TABLE A.6 Discrete State Space Block Properties

Name	A	B	C	D	X0	Sample Time
Servo Compensator	Acon_d	Bcon_d	eye(size(Acon_d))	Dcon_d	0	Tsamp

TABLE A.7 Display Block Properties

Name	Format	Decimation	Floating	SampleTime
THD Van	Short	1	off	−1
Van RMS	Short	1	off	−1
Vbn RMS	Short	1	off	−1
Vcn RMS	Short	1	off	−1

TABLE A.8 Fcn Block Properties

Name	Expr
Fcn1	sqrt(u[1]*u[1]+u[2]*u[2])
Fcn2	u[3]/sqrt(u[1]*u[1]+u[2]*u[2])
Fcn1	sqrt(u[1]*u[1]+u[2]*u[2])
Fcn2	Ilimit/sqrt(u[1]*u[1]+u[2]*u[2])

TABLE A.9   From Block Properties

Name	Goto Tag
From1	Vpwm_qd
From10	Vload_d
From11	Vref_d
From12	Iload_qd0
From13	Vload_ABC
From14	Vload_ABC
From15	Iload_qd0
From16	Iinv_qd
From17	Iinv_qd
From2	Vpwm_qd
From30	Vref_qd
From4	States
From5	States
From8	Vload_q
From9	Vref_q

TABLE A.10   Goto Block Properties

Name	Goto Tag	Tag Visibility
Goto	States	Local
Goto1	Vload_q	Local
Goto10	Itrans_qd	Local
Goto14	Vref_qd	Local
Goto2	Iload_qd0	Local
Goto20	Vinv_q	Local
Goto21	Vinv_d	Local
Goto3	Vref_q	Local
Goto4	Vref_d	Local
Goto5	Vload_d	Local
Goto6	Vpwm_qd	Local
Goto8	Vload_ABC	Local
Goto9	Iinv_qd	Local

TABLE A.11   Mux Block Properties

Name	Inputs	Display Option
Mux1	2	Bar
Mux2	2	Bar
Mux3	3	Bar
Mux6	2	Bar
Mux7	2	Bar
Mux6	2	Bar
Mux7	2	Bar

TABLE A.12   Product Block Properties

Name	Inputs	Saturate On Integer Overflow
Product	2	On
Product	2	On

TABLE A.13   Relational Operator Block Properties

Name	Operator
Relational operator	<
Relational operator	<=
Relational operator	<=

TABLE A.14   Selector Block Properties

Name	Elements	Input Port Width
D Output voltage	[6]	13
DQ Inv curr	[3:4]	13
DQ Inv voltages	[1:2]	13
DQ Output voltages 2	[5:7]	13
DQ Transformer curr 1	[8:9]	13
Load currents	[11:13]	13
Measurable states	[1:6 11:12]	13
Q Output voltage	[5]	13
Iinv	[3:4]	8
Iload	[7:8]	8
[Vinv Iinv]	[1:4]	8
Vload_qd	[5:6]	8

TABLE A.15   Sin Block Properties

Name	Amplitude	Frequency	Phase	Sample Time
D 5th harm	245.0*sqrt(2)*0.0	2*pi*ffun*7	−pi/2*7	0
D 7th harm	245.0*sqrt(2)*0	2*pi*ffun*3	pi/2*3	0
D Fund	120*sqrt(2)	2*pi*ffun	pi/2	0
Q 5th harm	245.0*sqrt(2)*0.0	2*pi*ffun*7	0	0
Q 7th harm	245.0*sqrt(2)*0	2*pi*ffun*3	0	0
Q Fundamental	120*sqrt(2)	2*pi*ffun	0	0

TABLE A.16   State Space Block Properties

Name	A	B	C	D	X0
Plant	Afl	Bsim	Cfl	Dsim	0

TABLE A.17   Sum Block Properties

Name	Icon Shape	Inputs	Saturate on Integer Overflow
Sum2	Rectangular	+-+	On
Sum3	Rectangular	+-+	On
Sum	Rectangular	+--	On
Sum	Rectangular	+-	On

TABLE A.18   Switch Block Properties

Name	Threshold
Switch	0.5
Switch	0.5
Switch	0.5

TABLE A.19   Unit Delay Block Properties

Name	X0	Sample Time
Unit Delay1	0	Tsamp
Unit Delay2	0	Tsamp

TABLE A.20   Zero-Order Hold Block Properties

Name	Sample Time
Zero-Order Hold5	Tsamp
Zero-Order Hold6	Tsamp
Zero-Order Hold7	Tsamp
Zero-Order Hold8	Tsamp

- *dsimservo*
- Current Controller
- CBinv∗CA
- CBinv∗CE
- PWM command limit
- inv(K)5
- DQ0Stat2ABC2
- DQ0Stat2ABC3
- DQ0Stat2ABC4
- RMS
- RMS1
- RMS2
- Spectrum Analyzer
- Manual Switch
- UPS1 Vxn
- Voltage Controller
- Current Limit
- K

A.2   SYSTEM: CURRENT CONTROLLER

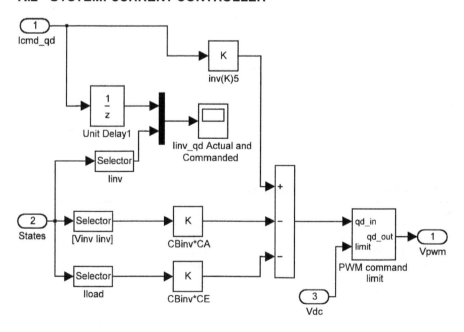

TABLE A.21   Current Controller System Information

Name	Current Controller	Parent	dsimservo
Description		Tag	
Blocks	Icmd_qd, States, Vdc, CBinv*CA, CBinv*CE, Iinv, Iinv_qd Actual and Commanded, Iload, Mux7, PWM command limit, Sum, Unit Delay1, [Vinv Iinv], inv(K)5, Vpwm	Link Status	None

A.3   SYSTEM: CURRENT LIMIT

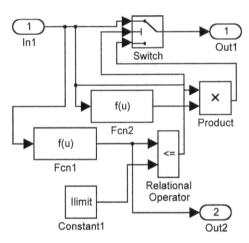

TABLE A.22   Current Limit System Information

Name	Current Limit	Parent	dsimservo/Voltage Controller
Description		Tag	
Blocks	In1, Constant1, Fcn1, Fcn2, Product, Relational Operator, Switch, Out1, Out2	Link Status	None

A.4 SYSTEM: DSIMSERVO

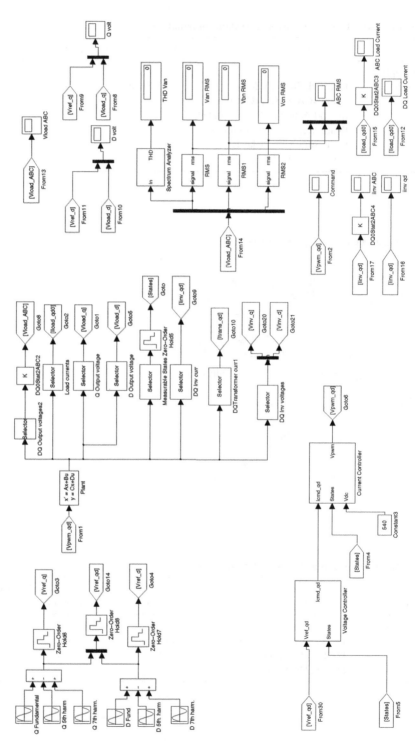

TABLE A.23   dsimservo System Information

Name	dsimservo	Parent	\<root\>
Description		Tag	
Blocks	ABC Load Current, ABC RMS, Command, Constant3, Current Controller, D 5th harm, D 7th harm, D Fund, D Output voltage, D volt, DQ Inv curr, DQ Inv voltages, DQ Load Current, DQ Output voltages2, DQ0Stat2ABC2, DQ0Stat2ABC3, DQ0Stat2ABC4, DQTransformer curr1, Demux, Demux2, From1, From10, From11, From12, From13, From14, From15, From16, From17, From2, From30, From4, From5, From8, From9, Goto, Goto1, Goto10, Goto14, Goto2, Goto20, Goto21, Goto3, Goto4, Goto5, Goto6, Goto8, Goto9, Load currents, Measurable States, Mux1, Mux2, Mux3, Mux6, Plant, Q 5th harm, Q 7th harm, Q Fundamental, Q Output voltage, Q volt, RMS, RMS1, RMS2, Spectrum Analyzer, Sum2, Sum3, THD Van, Van RMS, Vbn RMS, Vcn RMS, Vload ABC, Voltage Controller, Zero-Order Hold5, Zero-Order Hold6, Zero-Order Hold7, Zero-Order Hold8, iinv ABC, iinv qd	Link Status	

A.5 SYSTEM: PWM COMMAND LIMIT

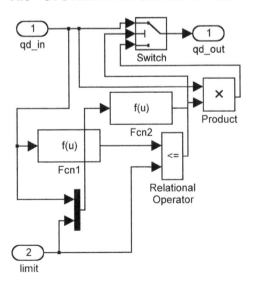

TABLE A.24   PWM command limit System Information

Name	PWM Command Limit	Parent	dsimservo/Current Controller
Description		Tag	
Blocks	qd_in, limit, Fcn1, Fcn2, Mux7, Product, Relational Operator, Switch, qd_out	Link Status	None

A.6 SYSTEM: SPECTRUM ANALYZER

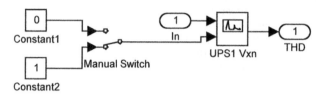

TABLE A.25   Spectrum Analyzer System Information

Name	Spectrum Analyzer	Parent	dsimservo
Description		Tag	
Blocks	In, Constant1, Constant2, Manual Switch, UPS1 Vxn, THD	Link Status	None

A.7 SYSTEM: VOLTAGE CONTROLLER

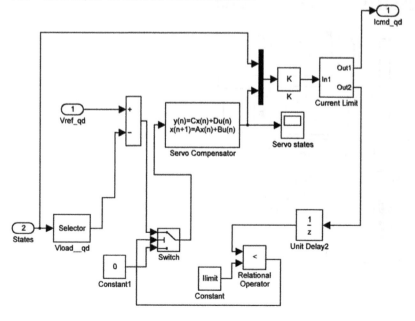

TABLE A.26   Voltage Controller System Information

Name	Voltage Controller		Parent	dsimservo
Description		Tag		
Blocks	Vref_qd, States, Constant, Constant1, Current Limit, K, Mux6, Relational Operator, Servo Compensator, Servo states, Sum, Switch, Unit Delay2, Vload__qd, Icmd_qd	Link Status	None	

A.8   SIMULATION TESTBED FOR CHAPTER 3

A.8.1   MATLAB Program: dsg.m

```
%%%%%%%%%%%%%%%%%%%%%%%%%%%%%%%%%%%%%%%%%%%%%%%%%%%%%%%%%%%%%%%%%%%%%%%%%%%%
%%  Program to calculate Control Gains using RSP Voltage Controller and   %%
%%  discrete Sliding Mode Current Controller for Single Unit Inverter     %%
%%%%%%%%%%%%%%%%%%%%%%%%%%%%%%%%%%%%%%%%%%%%%%%%%%%%%%%%%%%%%%%%%%%%%%%%%%%%

clear all

hdroop_enable=1;
m1 = 2e-6;
m2 = 10e-6;
n1 = 2e-4;
n2 = 20e-4;

Pi = 0.5*600e3*0.8/3;
Qi = 0.5*600e3*0.6/3;

% PWM Inverter Voltage Rating and Filter Components
KVA=600e3;
inv_volt=245*4;                     % L-L Inverter Cap Voltage
out_volt=208*4;                     % L-L output voltage
tr=out_volt/sqrt(3)/inv_volt;   % transformer turn ratio
Cinv=300e-6;                        % Inverter capacitor filter
Linv=5e-3;                      % Inverter inductor filter
Cload=90e-6;                    % % Output capacitor

% Define control sampling time
Tsamp=1/60/90;

% Define fundamental frequency
ffun=60;
wfun=2*pi*ffun;

%% Define harmonics frequencies
w1=wfun;        % 1st harmonic
w3=3*wfun;      % 3rd harmonic
w5=5*wfun;      % 5th harmonic
w7=7*wfun;      % 7th harmonic
```

```
%% Define KVA, voltages, currents, and impedance bases
KVAb=1/3*KVA;              % Single phase KVA base
Vb_inv=245*4/sqrt(3);      % Inv Filter voltage base
Ib_inv=KVAb/Vb_inv;        % Inv Filter current base
Vb_load=208*4/sqrt(3);     % Load voltage base
Ib_load=KVAb/Vb_load;      % Load current base
Zb_inv=Vb_inv/Ib_inv;      % Inverter Filter impedance base
Zb_load=Vb_load/Ib_load;   % Load impedance base

%% Compute perunit value of filter components
xCinv=1/(wfun*Cinv)/Zb_inv   % Inverter capacitor filter
xLinv=(wfun*Linv)/Zb_inv     % Inverter inductor filter
xCload=1/(wfun*Cload)/Zb_load  % Output capacitor
xLtrans=0.03                 % 3% p.u transformer inductance

%% Per Unit Current Limit
Ilimit=3*sqrt(2)*KVA/480;            % 300% inverter current limit

%% Define Delta-Wye Transformer Voltages and Currents Transfer Matrices
%In ABC reference frame
Tri=[1 -2 1;
     1  1 -2;
    -2  1 1];
Trv=[0  0 -1;
    -1  0  0;
     0 -1  0];

%In DQ reference frame
Tri_qd=3/2*tr*[1 1/sqrt(3) ;
      -1/sqrt(3) 1 ];
Trv_qd=3/2*tr*[1 -1/sqrt(3);
    1/sqrt(3) 1 ];

%In DQ0 reference frame
Tri_qd0=3/2*tr*[1 sqrt(3) 0;
      -sqrt(3) 1 0];
Trv_qd0=1/2*tr*[1 -sqrt(3);
       sqrt(3) 1 ;
       0    0 ];

%% Useful matrices
%ABC to DQ Stationary or vice versa
Ks=2/3*[cos(0) cos(0-2*pi/3) cos(0+2*pi/3);
    sin(0) sin(0-2*pi/3) sin(0+2*pi/3);
     1         1            1    ];
Ksinv=inv(Ks)  ;

%% DQ Line-to-line to Line To Neutral
Kln_qd=1/2*[1 -1/sqrt(3);
    1/sqrt(3) 1 ];

%%%%%%%%%%%%%%%%%%%%%%%%%%%%%%%%%%%%%%%%%%%%%%%%%%%%%%%%%%%%%%%%%%%%%%%%%%%%
% Start design of discrete SM current controller
%%%%%%%%%%%%%%%%%%%%%%%%%%%%%%%%%%%%%%%%%%%%%%%%%%%%%%%%%%%%%%%%%%%%%%%%%%%%

%% Define the plant for the current controller
A=[0                      xCinv*wfun/3 ;
   -1/(xLinv/wfun)           0];
```

```
B=[0;
   1/(xLinv/wfun)];
E=[-xCinv*wfun/(3);
   0];
F=[0;
   -1];
C=[0 1];
D=0;

%% Discretize the plant for the current controller
sysc=ss(A,B,C,zeros(size(C,1),size(B,2)),'inputdelay',0);
sysd=c2d(sysc,Tsamp,'zoh');
[Acurrd,Bcurrd,Ccurrd,Dcurrd]=ssdata(sysd);
CBinv=inv(Ccurrd*Bcurrd);
CA=Ccurrd*Acurrd;
CD=Ccurrd*F;
sysc=ss(A,E,C,zeros(size(C,1),size(B,2)),'inputdelay',0);
sysd=c2d(sysc,Tsamp,'zoh');
[Acurrd1,Ecurrd,Ccurrd1,Dcurrd1]=ssdata(sysd);
CE=Ccurrd1*Ecurrd;

%%% Discrete sliding mode controllers gains
Ksm_q=[CBinv(1,1)  -CBinv(1,1)*CA(1,1)*[1.5 -0.5]
      -CBinv(1,1)*CA(1,2)*[1.5 -0.5] -CBinv(1,1)*CE(1,1)*[1.5 -0.5] ];


%%%%%%%%%%%%%%%%%%%%%%%%%%%%%%%%%%%%%%%%%%%%%%%%%%%%%%%%%%%%%%%%%%%%%%%%%%%%%
% Start design of Perfect RSP voltages controllers
%%%%%%%%%%%%%%%%%%%%%%%%%%%%%%%%%%%%%%%%%%%%%%%%%%%%%%%%%%%%%%%%%%%%%%%%%%%%%


%% Define the true plant
Ao=[   0          xCinv*wfun/3  0      -xCinv*wfun/3;
     -1/(xLinv/wfun)       0        0      0;
       0            0        0    xCload*wfun;
     1/(xLtrans/wfun)      0    -1/(xLtrans/wfun)   0];

Bo=[0;
    1/(xLinv/wfun);
    0;
    0];

Co=[0 0 1 0];

%% Define the analog servo compensator
Ch1=[0  1;
    -w1^2 0];

Ch3=[0  1;
    -w3^2 0];

Ch5=[0  1;
    -w5^2 0];

Ch7=[0  1;
    -w7^2 0];
```

```
Ch_star=[[Ch1            zeros(2,2)   zeros(2,2)   zeros(2,2)   ];
         [zeros(2,2)      Ch3          zeros(2,2)   zeros(2,2)   ];
         [zeros(2,2)      zeros(2,2)   Ch5          zeros(2,2)   ]
         [zeros(2,2)      zeros(2,2)   zeros(2,2)   Ch7  ]];

Bh_star=[0;
         1;
         0;
         1;
         0;
         1;
         0;
         1];

% Discretize true plant
sysc=ss(Ao,Bo,Co,zeros(size(C,1),size(B,2)),'inputdelay',0.5*Tsamp);
sysd=c2d(sysc,Tsamp,'zoh');
[Aod,Bod,Cd,Dd]=ssdata(sysd);

% Calculate equivalent plant+DSM current controller
C1=[eye(2)  zeros(2,3)];
C2=[zeros(1,3)  1 0];
Ad=Aod-Bod*(CBinv*CA*C1+CBinv*CE*C2);
Bd=Bod*CBinv;
Cc_star=eye(size(Ch_star,1));

% Discetize and compute balanced realization of the controller
csysc=ss(Ch_star,Bh_star,Cc_star,zeros(size(Ch_star,1),size(Bh_star,2)));
[csysbc,Tbal]=ssbal(csysc);
csysd=c2d(csysbc,Tsamp,'zoh');
[Acon_d,Bcon_d,Ccon_d,Dcon_d]=ssdata(csysd);

% Form the augmented equivalent plant and the servo compensator
Ad_big=[Ad    zeros(size(Ad,1),size(Acon_d,2));
        -Bcon_d*Cd            Acon_d];
Bd_big=[Bd ; -Bcon_d*Dd];

% Define the weighting matrices
epsilon=1e-5;
Q2=eye(size(Acon_d,1));
%state_W=0.005 ;
state_W=0.6 ;
Q1=state_W*eye(size(Ad,1));
Q2(3:8,:)=0.1*Q2(3:8,:);
Q2(3:8,:)=0.1*Q2(3:8,:);

Q=[    Q1                    zeros(size(Ad,1),size(Acon_d,2));
   zeros(size(Acon_d,1),size(Ad,2))              5.5e5*Q2];

R=epsilon;

% Now perform the optimal calculations of the gains
[Kd,S,E]=dlqr(Ad_big,Bd_big,Q,R);
Kd=-Kd;

% Scale the inputs by 300 to make the Bcon_d matrix nice
%Bcon_d=300*Bcon_d;
%Kd(:,6:13)=Kd(:,6:13)/300;
```

```
Bcon_d=4*300*Bcon_d;
Kd(:,6:13)=Kd(:,6:13)/(4*300);

%%%%%%%%%%%%%%%%%%%%%%%%%%%%%%%%%%%%%%%%%%%%%%%%%%%%%
%% Compute Variable Frequency Harmonic Servo Compensator
%%%%%%%%%%%%%%%%%%%%%%%%%%%%%%%%%%%%%%%%%%%%%%%%%%%%%

harmonic=[1 3 5 7];
Bh_star=[0;1];

for h=1:length(harmonic)
    i=1;
    for f=59.0:0.02:61
      w=2*pi*f*harmonic(h);
        Ch_star=[0        1;
                 -w^2    0];
        csysc=ss(Ch_star,Bh_star,eye(2),0);
        [csysbc,T]=ssbal(csysc);
        csysd=c2d(csysbc,Tsamp,'zoh');
        [Acon_dx,Bcon_dx,Ccon_dx,Dcon_dx]=ssdata(csysd);
        A11(i)=Acon_dx(1,1);
        A12(i)=Acon_dx(1,2);
        A21(i)=Acon_dx(2,1);
        A22(i)=Acon_dx(2,2);
        B1(i)=Bcon_dx(1)*300;
        B2(i)=Bcon_dx(2)*300;
        i=i+1;
    end

    f=59.0:0.02:61;
    pA11{h} = polyfit(f,A11,1);
    pA12{h} = polyfit(f,A12,1);
    pA21{h} = polyfit(f,A21,1);
    pA22{h} = polyfit(f,A22,1);
    pB1{h}=polyfit(f,B1,1);
    pB2{h}=polyfit(f,B2,1);

    pA11{h}(2)=pA11{h}(2)+pA11{h}(1)*60;
    pA12{h}(2)=pA12{h}(2)+pA12{h}(1)*60;
    pA21{h}(2)=pA21{h}(2)+pA21{h}(1)*60;
    pA22{h}(2)=pA22{h}(2)+pA22{h}(1)*60;
    pB1{h}(2)=pB1{h}(2)+pB1{h}(1)*60;
    pB2{h}(2)=pB2{h}(2)+pB2{h}(1)*60;
end
```

A.8.2 MATLAB Program: plot.m

```
% Inputs
t= y(:,1);
P1= y(:,2);
Q1= y(:,3);
P2= y(:,4);
Q2= y(:,5);
I1a= y(:,6);
I1b= y(:,7);
```

```
I1c= y(:,8);
I2a= y(:,9);
I2b= y(:,10);
I2c= y(:,11);
Ia= y(:,12);
Ib= y(:,13);
Ic= y(:,14);
Va= y(:,15);
Vb= y(:,16);
Vc=y(:,17);


figure(1)
plot(t, I1a,'-', t, I1b,'-',t, I1c,'-',t, I2a,'-.',t, I2b,'-.',t, I2c,'-.')
%plot(t, Iload1,'-', t, Iload2,'-.')
grid
xlabel('Time [sec]')
ylabel('I_L_1, I_L_2 [A]')
%title('Load Voltage / Total Load Current')
axis([1.5 1.6 -800 800])
%legend('V_L','I_L')

%ylabel('I_L_1, I_L_2 [A]')
title('Two Load Currents (Transient)')
%axis([5.9 6.5 -1300 1300])
%legend('I_L_1','I_L_2')

figure(2)
plot(t, I1a,'-', t, I2a,'-')
grid
xlabel('Time [sec]')
ylabel('I_L_1_a, I_L_2_a [A]')
title('Two Load Currents')
axis([1.5 1.6 -800 800])
%axis([9.9 10 -1000 1000])
legend('I_L_1_a','I_L_2_a')

figure(3)
plot(t, P1/1000,'-', t, P2/1000,'-.')
grid
xlabel('Time [sec]')
ylabel('P_1, P_2 [kW]')
title('Two Active Powers')
axis([0 1.6 0 500])
%axis([0 10 0 40e3])
legend('P_1','P_2')

figure(4)
%subplot(2,2,4)
plot(t, Q1/1000,'-', t, Q2/1000,'-.')
grid
xlabel('Time [sec]')
ylabel('Q_1, Q_2 [kvar]')
title('Two Reactive Powers')
axis([0 1.6  0 500])
%axis([0 10 0 40e3])
legend('Q_1','Q_2')
```

APPENDIX B

FILE SSMODE.M

B.1 CHAPTERS 6, 7, AND 8: MATLAB CODE FOR ANALYSIS

B.1.1 Frequency Domain Analysis

```
%%%%%%%%%%%%%%%%%%%%%%%%%%%%%%%%%%%%%%%%%%%%%%%%%%%%%%%%%%%%%%%%%%%%%%%%%%%%
% Frequency domain ananlysis of 3-ph 4-wire transformerless inverter
% control using RSC and DSMC. Frequency responses of the open-loop plant,
% closed current loop, and closed voltage loop with nominal parameters are
% calculated.
% Min Dai 06/16/2005
%%%%%%%%%%%%%%%%%%%%%%%%%%%%%%%%%%%%%%%%%%%%%%%%%%%%%%%%%%%%%%%%%%%%%%%%%%%%

clear all

% PWM Inverter Voltage Rating and Filter Components
KVA=5e3;
inv_volt=240/sqrt(3);          % L-L Inverter Cap Voltage
out_volt=120;                  % L-N output voltage
tr=out_volt/inv_volt;          % transformer turn ratio
Cinv=55e-6;                    % Inverter filter capacitor
Linv=10.2e-3;                  % Inverter filter inductor = 1.8+2*4.2
                               % 3 reactors connected in series
                               % RMS current <= 12 Amperes
Rinv=1;                        % many inductors connected in series
```

Integration of Green and Renewable Energy in Electric Power Systems. By A. Keyhani, M. N. Marwali, and M. Dai
Copyright © 2010 John Wiley & Sons, Inc.

```
fs=40e6/2/3704;                    % PWM switching frequency

% Define control sampling time
Tsamp=1/fs;
%Tsamp=1e-6*fix(Tsamp*1e6);

% Define fundamental frequency
ffun=60;
wfun=2*pi*ffun;

%% Define harmonics frequencies
w1=wfun;        % 1st harmonic
w2=2*wfun;      % 2rd harmonic
w3=3*wfun;      % 3rd harmonic
w4=4*wfun;      % 4rd harmonic
w5=5*wfun;      % 5th harmonic
w7=7*wfun;      % 7th harmonic

%% Define KVA, voltages, currents, and impedance bases
KVAb=1/3*5e3;                       % Single phase KVA base
Vb_load=out_volt*sqrt(2);           % Load voltage base
Ib_load=KVAb/out_volt*sqrt(2);      % Load current base
Zb_load=Vb_load/Ib_load;            % Load impedance base


%% Compute perunit value of filter components
xCinv=1/(wfun*Cinv)/Zb_load        % Inverter capacitor filter
xLinv=(wfun*Linv)/Zb_load          % Inverter inductor filter
xRinv=Rinv/Zb_load;

%% Per Unit Current Limit
Ilimit=3;           % 300% inverter current limit


%%%%%%%%%%%%%%%%%%%%%%%%%%%%%%%%%%%%%%%%%%%%%%%%%%%%%%%%%%%%%%%%%%%%%
% Start design of discrete SM current controller
%%%%%%%%%%%%%%%%%%%%%%%%%%%%%%%%%%%%%%%%%%%%%%%%%%%%%%%%%%%%%%%%%%%%%

%% Define the plant for the current controller

A=[0                        xCinv*wfun ;
   -1/(xLinv/wfun)          -xRinv/(xLinv/wfun)];
B=[0;
   1/(xLinv/wfun)];
E=[-xCinv*wfun;
   0];
F=[0;
   -1];
C=[0 1];
D=0;

%% Discretize the plant for the current controller

sysc=ss(A,B,C,zeros(size(C,1),size(B,2)),'inputdelay',0);
sysd=c2d(sysc,Tsamp,'zoh');
```

```
[Acurrd,Bcurrd,Ccurrd,Dcurrd]=ssdata(sysd);
CBinv=inv(Ccurrd*Bcurrd);
CA=Ccurrd*Acurrd;
CD=Ccurrd*F;
sysc=ss(A,E,C,zeros(size(C,1),size(B,2)),'inputdelay',0);
sysd=c2d(sysc,Tsamp,'zoh');
[Acurrd1,Ecurrd,Ccurrd1,Dcurrd1]=ssdata(sysd);
CE=Ccurrd1*Ecurrd;

%%% Discrete sliding mode controllers gains
Ksm_q=[CBinv(1,1)  -CBinv(1,1)*CA(1,1)*[1.5 -0.5]  ...
       -CBinv(1,1)*CA(1,2)*[1.5 -0.5] -CBinv(1,1)*CE(1,1)*[1.5 -0.5] ];
%Ksm_q=[CBinv(1,1)  -CBinv(1,1)*CA(1,1)*[1 0]  ...
%       -CBinv(1,1)*CA(1,2)*[1 0] -CBinv(1,1)*CE(1,1)*[1 0] ];
                                               % no prediction


%% Define Discrete Sliding Mode Controller for frequency domain
%% analysis, see below.
%%   x=[vinv(k-1) iinv(k-1)  iload(k-1)]
%%   u=[icmd viinv iinv iload]
%%   y=vpwm

Asm=zeros(3,3);

Bsm=[0 1 0 0;
     0 0 1 0;
     0 0 0 1];

Csm=[Ksm_q(3) Ksm_q(5) Ksm_q(7)];

Dsm=[Ksm_q(1) Ksm_q(2) Ksm_q(4) Ksm_q(6)];

%%%%%%%%%%%%%%%%%%%%%%%%%%%%%%%%%%%%%%%%%%%%%%%%%%%%%%%%%%%%%%%%%%%
% Start design of Perfect RSP voltages controllers
%%%%%%%%%%%%%%%%%%%%%%%%%%%%%%%%%%%%%%%%%%%%%%%%%%%%%%%%%%%%%%%%%%%

%% Define the true plant

Ao=A;

Bo=B;

Co=[1 0];

Do=0;

%% Define the analog servo compensator
Ch1=[0  1;
     -w1^2 0];

Ch2=[0  1;
     -w2^2 0];

Ch3=[0  1;
     -w3^2 0];
```

```
Ch4=[0  1;
     -w4^2 0];

Ch5=[0  1;
     -w5^2 0];

Ch7=[0  1;
     -w7^2 0];


Ch_star=[[Ch1          zeros(2,2)   zeros(2,2)   zeros(2,2) ];
         [zeros(2,2)    Ch3          zeros(2,2)   zeros(2,2) ];
         [zeros(2,2)    zeros(2,2)   Ch5          zeros(2,2) ];
         [zeros(2,2)    zeros(2,2)   zeros(2,2)   Ch7        ]
        ];

Bh_star=[0;
         1;
         0;
         1;
         0;
         1;
         0;
         1;];

% Discretize true plant
sysc=ss(Ao,Bo,Co,Do,'inputdelay',0.5*Tsamp);
sysd=c2d(sysc,Tsamp,'zoh');
[Aod,Bod,Cd,Dd]=ssdata(sysd);

C1=[1 0 0;
    0 1 0   ];
Ad=Aod-Bod*CBinv*CA*C1;
Bd=Bod*CBinv;

Cc_star=eye(size(Ch_star,1));
% Discetize and compute balanced realization of the controller
csysc=ss(Ch_star,Bh_star,Cc_star,zeros(size(Ch_star,1),size(Bh_star,2)))
;
[csysbc,Tbal]=ssbal(csysc);
csysd=c2d(csysbc,Tsamp,'zoh');
[Acon_d,Bcon_d,Ccon_d,Dcon_d]=ssdata(csysd);


% Form the augmented equivalent plant and the servo compensator
Ad_big=[Ad   zeros(size(Ad,1),size(Acon_d,2));
        -Bcon_d*Cd          Acon_d];
Bd_big=[Bd ; -Bcon_d*Dd];

% Define the weighting matrices
epsilon=1e-5;
% state_W=0.005;        WP
% fund_servo_W=3e5;   5e5;   %WS1
% harm_servo_W=1e-3*1e4;   WH

state_W=0.005;        %WP
```

```
fund_servo_W=5e5;    %5e5;      %WS1
harm_servo_W=5e5;    %WH


Q1=state_W*eye(size(Ad,1));

Q2=eye(size(Acon_d,1));
Q2(1:2,:)=fund_servo_W*Q2(1:2,:);
Q2(3:8,:)=harm_servo_W*Q2(3:8,:);


Q=[    Q1                      zeros(size(Ad,1),size(Acon_d,2));
   zeros(size(Acon_d,1),size(Ad,2))                 Q2];

R=epsilon;

% Now perform the optimal calculations of the gains
[Kd,S,E]=dlqr(Ad_big,Bd_big,Q,R);
Kd=-Kd;

% Frequency domain analysis for the 3-ph 4-wire inverter system
% (performed in discrete-time):

% LC filter characteristic, in real values
LC_sys=tf(1,[Linv*Cinv Rinv*Cinv 1]);
natural_freq=1/sqrt(Linv*Cinv)       % in rad/s
damping_ratio=Rinv*sqrt(Cinv/(4*Linv))

% parameter definition, in per unit
B=0.6;       % per-unit suseptance of load parallel inductor
YB=wfun*B;   % convert into complement of inductance
YG=0.8;      % load parallel resistor

Cnom=xCinv*wfun;
Lnom=xLinv/wfun;
Rnom=xRinv;

r_epsilon=0.01;  % per-unit series resistance in the load
inductive branch, negligible


Ap=[    -YG/Cnom     1/Cnom  -1/Cnom;
        -1/Lnom      -Rnom/Lnom  0;
%        YB           0          0   ];
         YB           0          -r_epsilon*YB   ];

Bp=[    0   ;
        1/Lnom  ;
        0        ];

Cp=[    1            0          0;
        0            1          0;
        YG           0          1   ];

Dp=[    0;
        0;
        0    ];
```

```
sysp=ss(Ap,Bp,Cp,Dp);                          % the plant

N1=500;
W1=logspace(2,4,N1);
[LCmag, LCphase]=bode(LC_sys,W1);
for i=1:N1,
    LCmagdb(i)=20*log10(LCmag(1,1,i));
    LCph(i)=LCphase(1,1,i);
end
figure(1)
%bode(sysp);                                    % open-loop plant bode plot
%bode(LC_sys);                                  % open-loop LC filter
bode plot
%grid;
subplot(2,1,1);
semilogx(W1, LCmagdb,'b-');grid;zoom on;
title('Bode Plot', 'FontName','times new roman', 'FontSize', 16);
ylabel('Magnitude (dB)', 'FontName','times new roman', 'FontSize', 16);
set(gca,'FontName','times new roman');
set(gca,'FontSize',16);

subplot(2,1,2);
semilogx(W1, LCph,'b-');grid;zoom on;
ylabel('Phase (deg)', 'FontName','times new roman', 'FontSize', 16);
set(gca,'FontName','times new roman');
set(gca,'FontSize',16);
xlabel('Frequency (rad/s)', 'FontName','times new roman', 'FontSize', 16);
% End of figure(1).

sysp.inputdelay=[0.5*Tsamp];
syspd=c2d(sysp,Tsamp,'zoh');                    % discrete time plant with
input time delay

%sysp_w=d2c(syspd,'tustin');                    % continuous plant with
input time delay
%[Ap_d,Bp_d,Cp_d,Dp_d]=ssdata(sysp_w);   % do it in continuous time
[Ap_d,Bp_d,Cp_d,Dp_d]=ssdata(syspd);     % do it in discrete-time


dsm=pck(Asm,Bsm,Csm,Dsm);                       % see above for the definitions
of the ABCD

Aservo=[    Acon_d      zeros(8,1);
            zeros(1,8)  0           ];

Bservo=[    Bcon_d      zeros(8,3);
            zeros(1,3)  1   ];

Cservo=[    Kd(4:11)    Kd(3)   ];

Dservo=[    0   Kd(1:2) 0   ];

servo=pck(Aservo,Bservo,Cservo,Dservo);
systemnames='dsm servo';
inputvar='[err; plant { 3 } ]';                 % plant{3} in-
cludes Vload, Iinv, Iload
```

```
outputvar='[dsm]';
input_to_dsm='[servo; plant(1,2,3)]';
input_to_servo='[err; plant(1:2) ; dsm]';
sysoutname='Hcontrol_d';
cleanupsysic = 'no';
sysic;


[Acontrol,Bcontrol,Ccontrol,Dcontrol]=unpck(Hcontrol_d);
sys_con_d=ss(Acontrol,Bcontrol,Ccontrol,Dcontrol,Tsamp);
%sys_con_c=d2c(sys_con_d,'tustin');                       % continuous controller
%[Acontrol,Bcontrol,Ccontrol,Dcontrol]=ssdata(sys_con_c);



plant_nom_w=pck(Ap_d,Bp_d,Cp_d,Dp_d);
controller_w=pck(Acontrol,Bcontrol,Ccontrol,Dcontrol);
systemnames='plant_nom_w controller_w';
inputvar='[vref]';
%outputvar='[vref-plant_nom_w(1)]';
outputvar='[plant_nom_w(1)]';
input_to_plant_nom_w='[controller_w]';
input_to_controller_w='[vref-plant_nom_w(1);plant_nom_w(1:3)]';
sysoutname='clp_w';
cleanupsysic = 'no';
sysic;

[Avlp,Bvlp,Cvlp,Dvlp]=unpck(clp_w);
%sys_vloop=ss(Avlp,Bvlp,Cvlp,Dvlp);
sys_vloop=ss(Avlp,Bvlp,Cvlp,Dvlp,Tsamp);

[Vmag, Vphase, W2]=bode(sys_vloop);
N2=length(W2)
for i=1:N2,
    Vmagdb(i)=20*log10(Vmag(1,1,i));
    Vph(i)=Vphase(1,1,i);
end
figure(2)
subplot(2,1,1);
semilogx(W2, Vmagdb,'b-');grid;zoom on;
axis([1e-3 1e5 -200 50]);
title('Bode Plot', 'FontName','times new roman', 'FontSize', 16);
set(gca,'FontName','times new roman');
set(gca,'FontSize',16);
ylabel('Magnitude (dB)', 'FontName','times new roman', 'FontSize', 16);

subplot(2,1,2);
semilogx(W2, Vph-360,'b-');grid;zoom on;
axis([1e-3 1e5 -180 180]);
ylabel('Phase (deg)', 'FontName','times new roman', 'FontSize', 16);
set(gca,'FontName','times new roman');
set(gca,'FontSize',16);
xlabel('Frequency (rad/s)', 'FontName','times new roman', 'FontSize', 16);
% End of figure(2).
figure(3)
pzmap(sys_vloop);
axis([-1 1 -1 1]);
```

```
grid                    % done with closed voltage loop analysis
title('Pole-Zero Map', 'FontName','times new roman', 'FontSize', 16);
ylabel('Imaginary Axis', 'FontName','times new roman', 'FontSize', 16);
set(gca,'FontName','times new roman');
set(gca,'FontSize',16);
xlabel('Real Axis', 'FontName','times new roman', 'FontSize', 16);

% Closed current loop (open voltage loop) analysis
systemnames='plant_nom_w dsm';
inputvar='[icmd]';
outputvar='[plant_nom_w(2)]';
input_to_plant_nom_w='[dsm]';
input_to_dsm='[icmd;plant_nom_w(1:3)]';
sysoutname='clp_i';
cleanupsysic = 'no';
sysic;

[Ailp,Bilp,Cilp,Dilp]=unpck(clp_i);
%sys_iloop=ss(Ailp,Bilp,Cilp,Dilp);
sys_iloop=ss(Ailp,Bilp,Cilp,Dilp,Tsamp);

[Imag, Iphase, W3]=bode(sys_iloop);
N3=length(W3)
for i=1:N3,
    Imagdb(i)=20*log10(Imag(1,1,i));
    Iph(i)=Iphase(1,1,i);
end
figure(4)
subplot(2,1,1);
semilogx(W3, Imagdb,'b-');grid;zoom on;
axis([1e-2 1e4 -10 10]);
title('Bode Plot', 'FontName','times new roman', 'FontSize', 16);
set(gca,'FontName','times new roman');
set(gca,'FontSize',16);
ylabel('Magnitude (dB)', 'FontName','times new roman', 'FontSize', 16);

subplot(2,1,2);
semilogx(W3, Iph,'b-');grid;zoom on;
axis([1e-2 1e4 -225 45]);
ylabel('Phase (deg)', 'FontName','times new roman', 'FontSize', 16);
set(gca,'FontName','times new roman');
set(gca,'FontSize',16);
xlabel('Frequency (rad/s)', 'FontName','times new roman', 'FontSize', 16);
% End of figure(4).
figure(5)
pzmap(sys_iloop);
grid                    % done with closed current loop analysis
title('Pole-Zero Map', 'FontName','times new roman', 'FontSize', 16);
ylabel('Imaginary Axis', 'FontName','times new roman', 'FontSize', 16);
set(gca,'FontName','times new roman');
set(gca,'FontSize',16);
xlabel('Real Axis', 'FontName','times new roman', 'FontSize', 16);
```

B.1.2 Robust Stability Analysis

```
%%%%%%%%%%%%%%%%%%%%%%%%%%%%%%%%%%%%%%%%%%%%%%%%%%%%%%%%%%%%%%%%%%%%%%%%%%%%
%%%%% Robust stability ananlysis of 3-ph 4-wire transformerless inverter
% control using RSC and DSMC
% Min Dai 06/12/2005
%%%%%%%%%%%%%%%%%%%%%%%%%%%%%%%%%%%%%%%%%%%%%%%%%%%%%%%%%%%%%%%%%%%%%%%%%%%%
%%%%

clear all

% PWM Inverter Voltage Rating and Filter Components
KVA=5e3;
inv_volt=240/sqrt(3);              % L-L Inverter Cap Voltage
out_volt=120;                   % L-N output voltage
tr=out_volt/inv_volt;           % transformer turn ratio
Cinv=55e-6;                      % Inverter filter capacitor
Linv=10.2e-3;                     % Inverter filter inductor = 1.8+2*4.2
                                  % 3 reactors connected in series
                                  % RMS current <= 12 Amperes
Rinv=1;                          % many inductors connected in series


fs=40e6/2/3704;                  % PWM switching frequency

% Define control sampling time
Tsamp=1/fs;
%Tsamp=1e-6*fix(Tsamp*1e6);

% Define fundamental frequency
ffun=60;
wfun=2*pi*ffun;

%% Define harmonics frequencies
w1=wfun;      % 1st harmonic
w2=2*wfun;    % 2rd harmonic
w3=3*wfun;    % 3rd harmonic
w4=4*wfun;    % 4rd harmonic
w5=5*wfun;    % 5th harmonic
w7=7*wfun;    % 7th harmonic

%% Define KVA, voltages, currents, and impedance bases
KVAb=1/3*5e3;                     % Single phase KVA base
Vb_load=out_volt*sqrt(2);        % Load voltage base
Ib_load=KVAb/out_volt*sqrt(2);   % Load current base
Zb_load=Vb_load/Ib_load;         % Load impedance base


%% Compute perunit value of filter components
%xCinv=1/(wfun*3*Cinv)/Zb_load    % Inverter capacitor filter
xCinv=1/(wfun*Cinv)/Zb_load          % Inverter capacitor filter
xLinv=(wfun*Linv)/Zb_load        % Inverter inductor filter
xRinv=Rinv/Zb_load;
```

```
%% Per Unit Current Limit
Ilimit=3;        % 300% inverter current limit

%%%%%%%%%%%%%%%%%%%%%%%%%%%%%%%%%%%%%%%%%%%%%%%%%%%%%%%%%%%%%%%%%%%%%%%%%
% Start design of discrete SM current controller
%%%%%%%%%%%%%%%%%%%%%%%%%%%%%%%%%%%%%%%%%%%%%%%%%%%%%%%%%%%%%%%%%%%%%%%%%

%% Define the plant for the current controller

A=[0                        xCinv*wfun ;
   -1/(xLinv/wfun)          -xRinv/(xLinv/wfun)];
B=[0;
   1/(xLinv/wfun)];
E=[-xCinv*wfun;
   0];
F=[0;
   -1];
C=[0 1];
D=0;

%% Discretize the plant for the current controller

sysc=ss(A,B,C,zeros(size(C,1),size(B,2)),'inputdelay',0);
sysd=c2d(sysc,Tsamp,'zoh');
[Acurrd,Bcurrd,Ccurrd,Dcurrd]=ssdata(sysd);
CBinv=inv(Ccurrd*Bcurrd);
CA=Ccurrd*Acurrd;
CD=Ccurrd*F;
sysc=ss(A,E,C,zeros(size(C,1),size(B,2)),'inputdelay',0);
sysd=c2d(sysc,Tsamp,'zoh');
[Acurrd1,Ecurrd,Ccurrd1,Dcurrd1]=ssdata(sysd);
CE=Ccurrd1*Ecurrd;

%%% Discrete sliding mode controllers gains
Ksm_q=[CBinv(1,1)  -CBinv(1,1)*CA(1,1)*[1.5 -0.5]   ...

        -CBinv(1,1)*CA(1,2)*[1.5 -0.5] -CBinv(1,1)*CE(1,1)*[1.5 -0.5] ];
%Ksm_q=[CBinv(1,1)  -CBinv(1,1)*CA(1,1)*[1 0]  ...
%    -CBinv(1,1)*CA(1,2)*[1 0] -CBinv(1,1)*CE(1,1)*[1 0] ]; % no prediction


%% Define Discrete Sliding Mode Controller for robust stability analysis,
%% see below.
%%  x=[vinv(k-1) iinv(k-1)  iload(k-1)]
%%  u=[icmd viinv iinv iload]
%%  y=vpwm

Asm=zeros(3,3);

Bsm=[0 1 0 0;
     0 0 1 0;
     0 0 0 1];

Csm=[Ksm_q(3) Ksm_q(5)  Ksm_q(7)];
```

```
Dsm=[Ksm_q(1) Ksm_q(2) Ksm_q(4) Ksm_q(6)];

%%%%%%%%%%%%%%%%%%%%%%%%%%%%%%%%%%%%%%%%%%%%%%%%%%%%%%%%%%%%%%%%%
% Start design of Perfect RSP voltages controllers
%%%%%%%%%%%%%%%%%%%%%%%%%%%%%%%%%%%%%%%%%%%%%%%%%%%%%%%%%%%%%%%%%

%% Define the true plant

Ao=A;

Bo=B;

Co=[1 0];

Do=0;

%% Define the analog servo compensator
Ch1=[0  1;
    -w1^2 0];

Ch2=[0  1;
    -w2^2 0];

Ch3=[0  1;
    -w3^2 0];

Ch4=[0  1;
    -w4^2 0];

Ch5=[0  1;
    -w5^2 0];

Ch7=[0  1;
    -w7^2 0];


Ch_star=[[Ch1          zeros(2,2)  zeros(2,2)  zeros(2,2) ];
        [zeros(2,2)  Ch3          zeros(2,2)  zeros(2,2) ];
        [zeros(2,2)  zeros(2,2)  Ch5          zeros(2,2) ];
        [zeros(2,2)  zeros(2,2)  zeros(2,2)  Ch7         ]
        ];

Bh_star=[0;
        1;
        0;
        1;
        0;
        1;
        0;
        1;];

% Discretize true plant
sysc=ss(Ao,Bo,Co,Do,'inputdelay',0.5*Tsamp);
sysd=c2d(sysc,Tsamp,'zoh');
[Aod,Bod,Cd,Dd]=ssdata(sysd);
```

```
% Calculate equivalent plant+DSM current controller
C1=[1 0 0;
    0 1 0   ];
Ad=Aod-Bod*CBinv*CA*C1;
Bd=Bod*CBinv;

Cc_star=eye(size(Ch_star,1));
% Discetize and compute balanced realization of the controller
csysc=ss(Ch_star,Bh_star,Cc_star,zeros(size(Ch_star,1),size(Bh_star,2)));
[csysbc,Tbal]=ssbal(csysc);
csysd=c2d(csysbc,Tsamp,'zoh');
[Acon_d,Bcon_d,Ccon_d,Dcon_d]=ssdata(csysd);


% Form the augmented equivalent plant and the servo compensator
Ad_big=[Ad      zeros(size(Ad,1),size(Acon_d,2));
        -Bcon_d*Cd          Acon_d];
Bd_big=[Bd ; -Bcon_d*Dd];

% Define the weighting matrices
epsilon=1e-5;

state_W=0.05;       %WP
fund_servo_W=5e7;   %5e5;    %WS1
harm_servo_W=5e5;   %WH


Q1=state_W*eye(size(Ad,1));

Q2=eye(size(Acon_d,1));
Q2(1:2,:)=fund_servo_W*Q2(1:2,:);
Q2(3:8,:)=harm_servo_W*Q2(3:8,:);


Q=[     Q1                      zeros(size(Ad,1),size(Acon_d,2));
    zeros(size(Acon_d,1),size(Ad,2))                    Q2];

R=epsilon;

% Now perform the optimal calculations of the gains
[Kd,S,E]=dlqr(Ad_big,Bd_big,Q,R);
Kd=-Kd;

% Robust stability analysis for the 3-ph 4-wire inverter system:

% parameter definition, in per unit
B=0.6;      % per-unit suseptance of load parallel inductor
YB=wfun*B;  % convert into complement of inductance
YG=0.8;     % load parallel resistor

Cnom=xCinv*wfun;
Lnom=xLinv/wfun;
Rnom=xRinv;

r_epsilon=0.01; % per-unit series resistance in the load induc-
tive branch, negligible
```

```
Ctol=0.06;
Ltol=0.15;
Rtol=0.5;
YBtol=1;
YGtol=1;

deltaR=xRinv*Rtol;
deltaYB=YB*YBtol;
deltaYG=YG*YGtol;

Ap=[    -YG/Cnom      1/Cnom  -1/Cnom;
        -1/Lnom       -Rnom/Lnom  0;
%        YB            0           0    ];
         YB            0          -r_epsilon*YB   ];

Bp=[   0    -Ctol  0      0             0         -deltaYG/Cnom;
       1/Lnom  0   -Ltol  -deltaR/Lnom  0         0;
       0       0   0      0             deltaYB   0    ];

Cp=[    1             0           0;
        0             1           0;
        YG            0           1;
       -YG/Cnom       1/Cnom     -1/Cnom;
       -1/Lnom        -Rnom/Lnom  0;
        0             1           0;
        1             0           0;
        1             0           0    ];

Dp=[    0      0    0    0            0   0;
        0      0    0    0            0   0;
        0      0    0    0            0   deltaYG;
        0     -Ctol 0    0            0   -deltaYG/Cnom;
        1/Lnom 0   -Ltol -deltaR/Lnom 0   0;
        0      0    0    0            0   0;
        0      0    0    0            0   0;
        0      0    0    0            0   0    ];

sysp=ss(Ap,Bp,Cp,Dp);                    % the plant
sysp.inputdelay=[0.5*Tsamp  zeros(1,5)];
syspd=c2d(sysp,Tsamp,'zoh');             % discrete time plant with
input time delay


sysp_w=d2c(syspd,'tustin');              % continuous plant with
input time delay
[Ap_c,Bp_c,Cp_c,Dp_c]=ssdata(sysp_w);

dsm=pck(Asm,Bsm,Csm,Dsm);                % see above for the
definitions of the ABCD

Aservo=[   Acon_d      zeros(8,1);
           zeros(1,8)  0            ];

Bservo=[   Bcon_d      zeros(8,3);
           zeros(1,3)  1    ];
```

```
Cservo=[    Kd(4:11)    Kd(3)    ];

Dservo=[    0    Kd(1:2) 0    ];

servo=pck(Aservo,Bservo,Cservo,Dservo);
systemnames='dsm servo';
inputvar='[err; plant{3}];        % plant{3} includes
Vload, Iinv, Iload
outputvar='[dsm]';
input_to_dsm='[servo; plant(1,2,3)]';
input_to_servo='[err; plant(1:2) ; dsm]';
sysoutname='Hcontrol_d';
cleanupsysic = 'yes';
sysic;


[Acontrol,Bcontrol,Ccontrol,Dcontrol]=unpck(Hcontrol_d);
sys_con_d=ss(Acontrol,Bcontrol,Ccontrol,Dcontrol,Tsamp);
sys_con_c=d2c(sys_con_d,'tustin');                    % continuous controller
[Acontrol,Bcontrol,Ccontrol,Dcontrol]=ssdata(sys_con_c);



plant_nom_w=pck(Ap_c,Bp_c,Cp_c,Dp_c);
controller_w=pck(Acontrol,Bcontrol,Ccontrol,Dcontrol);
systemnames='plant_nom_w controller_w';
inputvar='[vref ; w{5}]';
outputvar='[vref-plant_nom_w(1); plant_nom_w(4:8)]';
input_to_plant_nom_w='[controller_w; w(1:5)]';
input_to_controller_w='[vref-plant_nom_w(1);plant_nom_w(1:3)]';
sysoutname='clp_w';
cleanupsysic = 'yes';
sysic;

omega=logspace(-1,5,500);
clp_g=frsp(clp_w,omega);

clpstab_g=sel(clp_g,2:6,2:6);

figure(1)
for i=1:5
subplot(5,1,i)
vplot('liv,m',sel(clpstab_g,i,i));grid;zoom on;
ylabel(i, 'FontName','times new roman', 'FontSize', 16);
set(gca,'FontName','times new roman');
set(gca,'FontSize',16);
end
xlabel('Frequency (rad/s)', 'FontName','times new roman', 'FontSize', 16);

%deltaset = [-1 0;-1 0;-1 0; 1 1; 1 1];
deltaset = [-1 0;-1 0;-1 0; -1 0; -1 0];
[bnds,dvec,sens,pvec] = mu(clpstab_g,deltaset);
figure(2)
vplot('liv,m',bnds);
grid
pert=dypert(pvec,deltaset,bnds);
```

```
%% Compute closed loop poles zero for different delta_L
delta_C=0;
delta_R=0;
delta_YB=0;
delta_YG=0;
%close(3)
figure(3)
hold on
for delta_L=-[0 0.5 1.0 1.5 2.0 2.5]
delta=[ delta_C 0        0        0         0    ;
         0      delta_L 0        0         0    ;
         0      0       delta_R 0         0    ;
         0      0       0       delta_YB 0    ;
         0      0       0       0        delta_YG ];
clp_delta=starp(clp_w,delta);
[Adelta,Bdelta,Cdelta,Ddelta]=unpck(clp_delta);
sys_delta=ss(Adelta,Bdelta,Cdelta,Ddelta);
pzmap(sys_delta);
title('Pole-zero map', 'FontName','times new roman', 'FontSize', 16);
ylabel('Imaginary axis', 'FontName','times new roman', 'FontSize', 16);
xlabel('Real axis', 'FontName','times new roman', 'FontSize', 16);
set(gca,'FontName','times new roman');
set(gca,'FontSize',16);
end
```

B.2 CHAPTER 6: EXPERIMENTAL SETUP RECONFIGURATION

The basic configuration of the dual AC–DC–AC power converter units is described in a Technical Report entitled *Design and Implementation of Experimental Testbed for 5 KVA Uninterruptible Power Supply*, authored by Dr. M. N. Marwali and Profssor Keyhani, available at http://www.ece.osu.edu/ems/ with Serial Number TR036.

This chapter gives information about reconfiguring the setup on the three-phase four-wire system test and the grid-connected inverter test.

B.2.1 Reconfiguration for Three-Phase Four-Wire System Tests

Unit 1 (the left-hand side unit connected to host PCs EMS06 and EMS07) has been upgraded, allowing reconfiguration for the four-wire tests.

When a four-wire test is conducted, the power connections on the measurement box can be reconfigured as shown in Fig. B.1. The upgrading is on the rectifier side signal conditioning board: Three AC voltage measurement channels have been added (A, B, C, and N, four pins)—that is, popularized with components—to allow measurement of three-phase line-to-neutral voltage of the rectifier front end. The signal conditioning gain used is the "V240 Scale," the same as the "INPUT" channels. On the rectifier side, the three-phase voltage sensing signals are introduced out from the box through the four-pin "Bypass" connector and sent to DSP through the "120VAC" connector, which actually allows 240-V line-to-line voltage input. The current sensing signals (three-phase) come out from the box through "BYP A," "BYP B," and "BYP C"

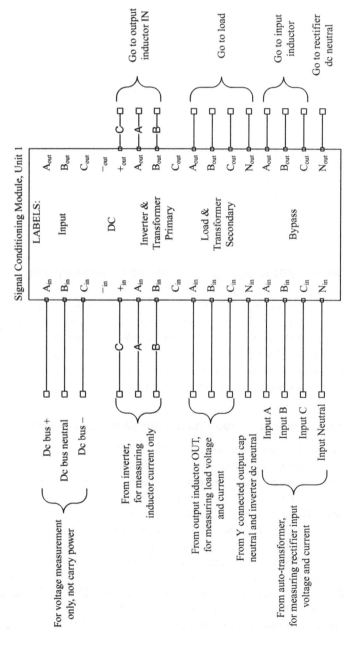

FIGURE B.1 The power connection reconfiguration of the measurement box of Unit 1 for three-phase four-wire split dc bus AC–DC–AC system tests.

connectors (two-pin each) and enter the DSP through the "I load A," "I load B," and "I load C" connectors. On the rectifier side, the current sensing connectors "INP A" and "INP B" (two-pin each) on the signal conditioning board are not used in the four-wire tests even though they are connected to the sensors inside the box.

One more voltage sensing connector labeled "INP (2)" (three-pin) has been added on the "Input" section of the box, combining the existing "INP" (three-pin) connector to send positive and negative halves of DC bus voltages to both the rectifier side signal conditioning board and the inverter side one. On the rectifier side board, the signals go to the "INP V" connector; and on the inverter side board, the signals go to the "INV V" connector.

One current sensing connector labeled "I_{DC}" (two-pin) has been introduced out from the "DC" section of the box, allowing current measurement on the "+in" to "+out" branch. This branch is used for measuring Phase C current of the inverter output filter inductor. This channel uses the "I Byp C" channel on the inverter side board to go to the DSP. Meanwhile, the sensor signals of Phases A and B of the inductor currents are introduced out from the box through the "Inv A" and "Inv B" connectors (two-pin each) and sent to DSP through the "I Inv A" and "I Inv B" channels.

The load voltage and current measurement follows the original configurations: Load voltage connector (four-pin) from the box goes to the "Load" connector on the board, and the load current connectors (three, two-pin each) go to the three load current connectors on the board.

None of the "V_{DC}" connectors (two-pin each) on either board is used in the four-wire tests.

B.2.2   Upgrading of Unit 2 for Grid-Connected Inverter Tests

Upgrading has been conducted on Unit 2 for grid-connected tests. Please refer to Fig. B.2 for the power stage connections on the measurement box. Notice that a second contactor has been installed inside the box for the "Bypass" section to perform the ON/OFF switch between the unit and the utility grid. For protection purpose, three fuses have been installed on the "Bypass" branches. All other power connections are the same as the original UPS unit.

Due to the limit number of ADC channels on the TI 2407 DSP chips (16 channels), measured bypass currents "I Byp B" and "I Byp C" on the inverter signal conditioning board are multiplexed to ADC channel 16. The DSP initialization sets the default ON signal to be "I Byp C." In the control program, ADC channel 16 is read twice each control cycle, one for "I Byp B" and the other or "I Byp C." Please refer to the code.

B.2.3   The Contactor and Its Drive Circuit

The contactor is Omron[®] G7J-3A1B-BW-1-AC10012. It is driven by the following control circuit, where the AQH solid-state optical relay is the key drive device:

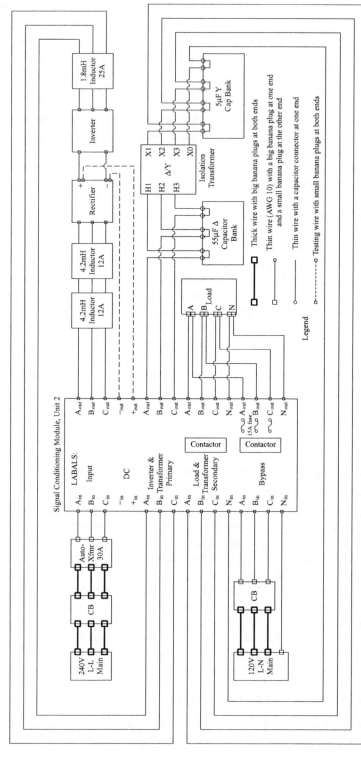

FIGURE B.2 The power connection reconfiguration of the measurement box Unit 2 for three-phase three-wire grid-connected inverter system tests.

1. Order the parts from http://www.alliedelec.com/
   2N2222A: NPN transistor
   KBP01G: diode rectifier bridge
   LM7805CT: volt regulator, not used
   17265: AC receptacle, not used
   AQH1223: opto coupler for contactor
   G7J-3A1B-BW-1-AC100120: The contactor
   SEK101M025ST: 100-μF 25-V electrolytic cap
   SW-210: 120-V/5-V transformer
   Resistors as shown in the circuit: $^1/_4$ watt, 1%
2. Buy the following parts from Radioshack:
   A project box
   A small general-purpose PCB board
   4 insulated PCB standoffs
   1 fuse holder: $1^1/_4 \times ^1/_4$
   1 250-V 250-mA fuse (fast)
   1 120-VAC lamp

B.2.4   The DC Bus Overvoltage Protection Board Design

A DC bus overvoltage protection board has been designed to provide independent hardware protection on the boosted dc bus besides the rectifier DSP software overvoltage protection for both three-wire and four-wire split DC bus protection. The protection circuit detects overvoltage on both top and bottom halves of the DC bus where the neutral of the DC bus is available from the connecting point of the two series-connected DC electrolytic capacitors supporting the DC voltage. Once overvoltage is detected on either half of the DC bus, two protective actions will be tripped, with the rectifier PWM shut down through PDPINTB and discharging of DC bus through a low induction regeneration resistor. The schematic of the protection circuit is shown below in Figs. B.4–B. B.6.

Figure B.4 shows the top and bottom half DC overvoltage detection circuit, including voltage divider, differential amplifier, and hysteresis comparator stages. The protection triggering thresholds are determined by the resistors used in these three stages. The hysteresis comparator is used to control the discharging branch to be ON when overvoltage is detected and OFF when the DC bus voltage drops below a certain voltage (190 V for half DC bus). Turning OFF the discharging will stop the large current flow through R (low-inductance power resistor: TL88K25R0 25Ω 114W) and let R1 and R2 (attached to the DC capacitors, 22k each) resume the discharging until the voltage drops to the uncontrolled rectifier level; for the 3-ph 240-V line-to-line input case, this voltage is $\sqrt{2} \times 240 = 340$ V for full DC bus, or say 170-V half DC bus. This is used to protect R from heating up by long-time

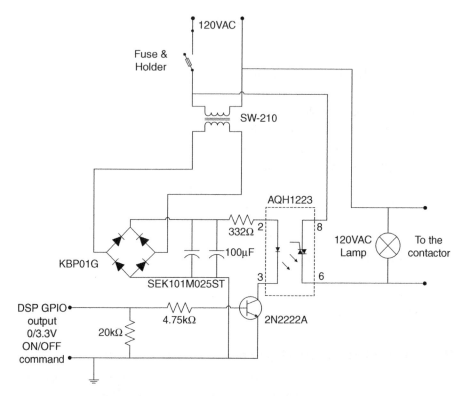

FIGURE B.3   The three-phase contactor drive circuit.

conducting, assuming that input AC voltage source is not removed. Figure B.5 shows on-board power management, opto-coupler isolation, and digital logic circuits. Figure B.6 shows board connectors. Figure B.7 shows how to connect the board to the power stage, where the error signals Error 1 to 3 are three BNC connectors on the Semikron 3-ph power converter teaching unit. This allows the board to also respond to the error messages sent by the power converter, which will also trigger the PDPINTB and the discharging. The gain and hysteresis designs for are given in the two MATLAB files: opamp_gain.m and opamp_calc.m.

There are two identical protection boards in total, one for each unit and sitting on top of the rectifier side signal conditioning board. In grid-connected operation or parallel operation of the units, if the PDPINTB signal on the protection board is introduced to the inverter side PDPINTB pin, the inverter PWM can be shut down at DC bus overvoltage caused by real power pumped back from the inverter.

```
% opamp_gain.m
% Dc bus overvoltage protection circuit gain calculation - the opamp
% Only the differential gain considered, no common mode voltage
% considered, for common mode voltage verification, see opamp_calc.m
```

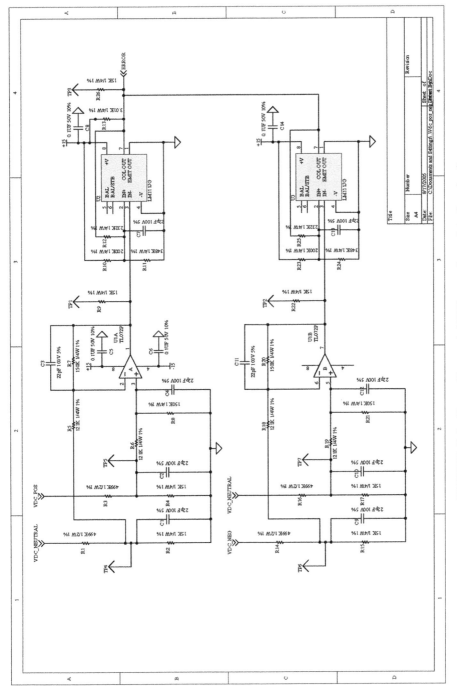

FIGURE B.4 Top and bottom half DC overvoltage detection circuit.

287

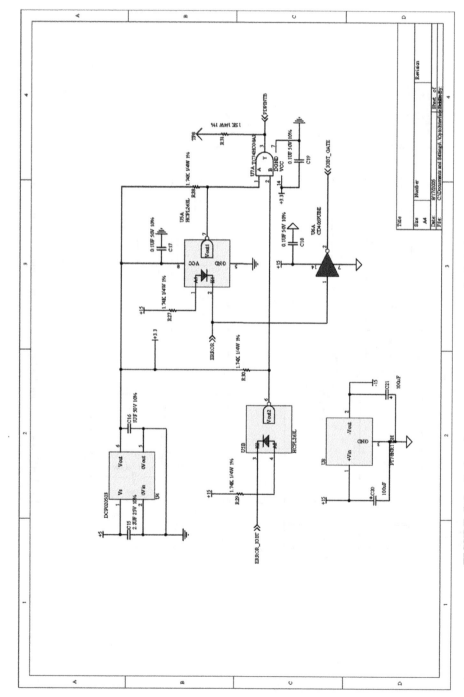

FIGURE B.5 On-board power management, opto-coupler isolation, and digital logic circuits.

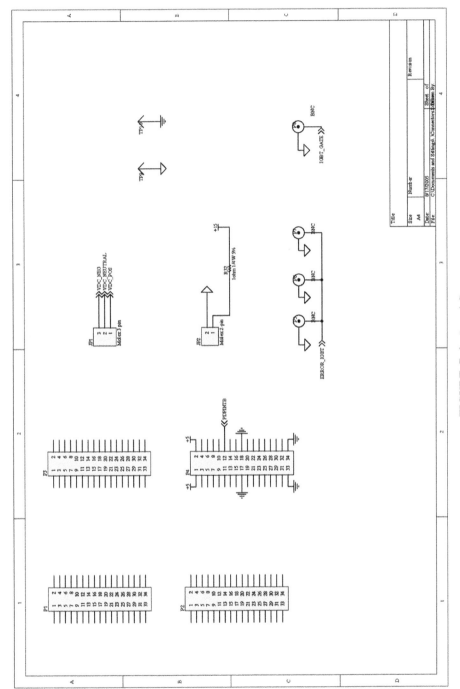

FIGURE B.6 Board Connectors.

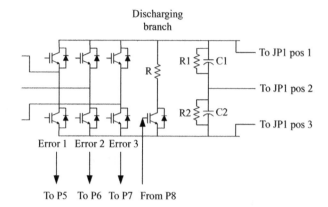

FIGURE B.7   Connecting the protection board to the power stage.

```
% Voltage divider and the OpAmp
R3=499e3;          % voltage divider top
R4=15e3;           % voltage divider bottom
R1=121e3;          % OpAmp input resistor
R2=150e3;          % OpAmp feedback and reference input resistor

Vin=170;

R34=R3*R4/(R3+R4);          % R3//R4

Vin_divided=R4/(R3+R4)*Vin
Vout=R4/(R3+R4)*R2/(R1+R34)*Vin
gain=R4/(R3+R4)*R2/(R1+R34)

gain_test=R2/R1     % ignore the voltage divider, test input signal hooked
                    % to the differential amplifier directly


Vin1=0.5*Vin_divided
Vin2=-0.5*Vin_divided
V_neg=Vin2-(Vin2-Vout)*(R1+R34)/(R1+R34+R2)     % OpAmp neg pin voltage
V_pos=R2*Vin1/(R1+R34+R2)                        % OpAmp pos pin voltage


R3_pwr=Vin^2/R3
R4_pwr=Vin_divided^2/R4

% The hysteresis comparator
Vin_th_hi=350;                      % protection threshold voltage, turn on
discharging path
Vin_th_lo=190;                      % turn off discharge path

Vth_exp_hi=Vin_th_hi*gain
Vth_exp_lo=Vin_th_lo*gain
```

```
R5=200e3;                    % comparator reference voltage divider top
R6=348e3;                    % comparator reference voltage divider bottom
R7=232e3;                  % comparator feedback voltage (to cause
hysteresis)
R8=3.01e3;                  % comparator collector out to Vcc resistor

R578=R5*(R7+R8)/(R5+R7+R8);      % R5//(R7+R8)
R67=R6*R7/(R6+R7);               % R6//R7

Vcc=15;
Vth_hi=R6/(R6+R578)*Vcc
Vth_lo=R67/(R5+R67)*Vcc

Vin_th_hi_cal=Vth_hi/gain
Vin_th_lo_cal=Vth_lo/gain

Vin_test_hi_cal=Vth_hi/gain_test
Vin_test_lo_cal=Vth_lo/gain_test

Vout_hi=(Vcc-Vth_hi)*R7/(R7+R8)+Vth_hi

R8_pwr=Vcc^2/R8


% opamp_calc.m
% Common mode input voltages are considered
% a complementary consideration on top of the differential-gain-only
% design in opamp_gain.m

format long

% protection circuit parameters
R1=499e3;
R2=15e3;
R3=121e3;
R4=150e3;
Vin_neg=-308;
Vin_pos=25;


R_pos=R2*(R3+R4)/(R2+R3+R4)  % R2//(R3+R4)


V_pos=Vin_pos*R_pos/(R1+R_pos)

V_opamp_pos=V_pos*R4/(R3+R4)

V_opamp_neg=V_opamp_pos       % virtual short

V_neg=(R1*R2*V_opamp_neg+R2*R3*Vin_neg)/(R1*R2+R2*R3+R3*R1)

Vout_calcA=(V_opamp_neg-V_neg)*R4/R3+V_opamp_neg

V_diff=V_pos-V_neg
```

```
gain_diff=R4/R3;
Vout_calcB=V_diff*gain_diff

gain=Vout_calcA/(Vin_pos-Vin_neg)
```

B.3 CHAPTER 7: ELECTRONIC FILES

Electronic files attached to this report include:

1. Simulink models of the three-phase four-wire AC–DC–AC system control on passive load and power flow of three-phase three-wire grid-connected inverter system plus associated initialization MATLAB M-files.
2. TI TMS320LF2407(A) DSP source code for three-phase four-wire AC–DC–AC system control. The source code for three-phase three-wire front-end rectifier control.

BIBLIOGRAPHY

http://www.electricitystorage.org/technologies_applic ations.htm.

http://www.electricitystorage.org/tech/technologies_technologies_flywheels.htm.

Min Dai, M.N. Marwali, Jin-Woo Jung, and A. Keyhani, Power flow control of a single distributed generation unit, *IEEE Trans. Power Electron.*, vol. *23*, issue 1, pp. 343–352, Jan. 2008.

Min Dai, M.N. Marwali, Jin-Woo Jung, and A. Keyhani, A three-phase four-wire inverter control technique for a single distributed generation unit in island mode, *IEEE Trans. Power Electron.*, vol. *23*, issue 1, Jan. 2008, pp. 322–331

Mohammad N. Marwali, Jin-Woo Jung and Ali Keyhani, Stability analysis of load sharing control for distributed generation systems, *IEEE Trans. Energy Conversion*, vol. *22*, No. 3, pp. 737–745, Sept. 2007.

Jin-Woo Jung and Ali Keyhani, Control of a fuel cell based Z-source converter, *IEEE Trans. Energy Conversion*, vol. *22*, no. 2, pp. 467–476, June 2007.

B. Kroposki, C. Pink, T. Basso, and R. DeBlasio, Microgrid standards and technology developmental, in *IEEE Power Engineering Society General Meeting*, 2007, Tampa, FL, 24–28 June 2007, pp. 1–4.

A. Malmedal, B. Kroposki, and P. Sen, Energy Policy Act of 2005 and Its Impact on Renewable Energy Applications in USA, in *IEEE Power Engineering Society General Meeting*, Tampa, FL, 24–28 June 2007, pp. 1–4.

C. Wang and M. H. Nehrir, Fuel cells and load transients, *IEEE Power Energy Mag.* vol. *5*, issue 1, pp. 58–63. Jan.–Feb. 2007,

*References appear in chronological order.

Integration of Green and Renewable Energy in Electric Power Systems. By A. Keyhani, M. N. Marwali, and M. Dai
Copyright © 2010 John Wiley & Sons, Inc.

G. Hernández-González and R. Iravani, Current injection for active islanding detection of electronically-interfaced distributed resources, *IEEE Trans. Power Delivery*, vol. *21*, no. 3, pp. 1698–1705, July 2006.

B. Kroposki J. Levene, K. Harrison, P. Sen, and F. Novachek, Electrolysis: opportunities for electric power utilities in a hydrogen economy, in *North American Power Symposium*, Carbondale, IL, September 2006, pp. 697–706.

B. Kroposki, C. Pink, R. DeBlasio, H. Thomas, M. Simoes, and P. Sen, Benefits of power electronic interfaces for distributed energy applications, *IEEE Power Engineering Society General Meeting*, Montreal, June 2006, pp.295–305.

B. Kroposki, T. Basso, and R. DeBlasio, and N. R. Friedman Interconnection of alternative energy sources with the grid, in *Invited Chapter in Integration of Alternative Sources of Energy*, F. Farret and M. Simoes, editors, John Wiley & Sons, Hoboken, NJ, 2006, pp. 295–305.

D. Menicucci, R. Ducey, and P. Volkman. Energy Surety for Mission Readiness. *Public Works Digest of the U.S. Army Installation Management Agency*, vol. *XVIII*, no. 2, A publication of the U.S. Army Installation Management Agency, March/April 2006.

R. S. Balog, Autonomous local control in distributed DC power systems, Ph.D. dissertation, University of Illinois at Urbana—Champaign, 2006.

C. Wang, M. H. Nehrir, and H. Gao, Control of PEM fuel cell distributed generation systems, *IEEE Trans. Energy Conversion*, vol. *21*, issue 2, pp. 586–595, June 2006.

S. Jemei, D. Hissel, A. Coince, and B. Al-Nasrawi, Optimization and economic analysis of an hybrid fuel cell, photovoltaic and battery electric power generation system, *J. Fuel Cell Sci. Technol.*, vol. *3–4*, pp. 410–414, 2006.

T. Abdallah, R. Ducey, R. S. Balog, C. Feickert, W. Weaver, A. Akhil, and D. Menicucci, *Control Dynamics of Adaptive and Scalable Power and Energy Systems for Military Microgrids*, U.S. Army Engineer Research and Development Center, 2006.

E. Gursoy and D. Niebur, On-line estimation of electric power system active loads neural networks, *International Joint Conference on Neural Networks*, 16–21 July 2006 pp. 1689–1694.

D. Menicucci, R. Ducey, and P. Volkman, Energy surety for mission readiness, *Public Works Digest of the U.S. Army Installation Management Agency*, vol. *18*, no. 2, A publication of the U.S. Army Installation Management Agency, March–April 2006.

R. H. Lasseter and P. Piagi, Control of small distributed energy resources, United States Patent 7116010, October 3, 2006.

E. C. Shaffer, D. D. Massie, and J. B. Cross, Power and energy architecture for army advanced energy initiative, Army Research Laboratory, 2006.

M. N. Marwali, M. Dai, and A. Keyhani, Robust stability analysis of voltage and current control for distributed generation systems, *IEEE Trans. Energy Conversion*, vol. *21*, no. 2, pp. 516–526, June 2006.

S. Jian, Input impedance analysis of single-phase PFC converters, *IEEE Trans. Power Electron.*, vol. *20*, no. 2, pp. 308–314, 2005.

R. Firestone and C. Marnay, Energy manager design for microgrids, Ernest Orlando Laurence Berkeley National Laboratory, LBNL-54447, January 2005.

C. Wang, M. H. Nehrir, and S. R Shaw, Dynamic models and model validation for PEM fuel cells using electrical circuits, *IEEE Trans. Energy Conversion*, vol. *20*, issue 2, pp. 442–451, June 2005.

M. Dai, M. N. Marwali, J.-W. Jung, and A. Keyhani, A PWM rectifier control technique for three-phase double conversion UPS under unbalanced load, *Applied Power Electronics Conference and Exposition*, vol. *1*, 2005, pp. 548–552.

F. Z. Peng, M. Shen, and Z. Qian, Maximum boost control of the Z-source inverter, *IEEE Trans. Power Electron.*, vol. *20*, pp. 833–838, 2005.

F. Z. Peng, M. Shen, and Z. Qian, State-feedback-with-integral control plus repetitive control for UPS inverters, in *Proceedings IEEE Applied Power Electronics Conference and Exposition (APEC '05)*, vol. *1*, Austin, TX, Mar. 2005, pp. 553–559.

J. W. Jung, M. Dai, and A. Keyhani, Optimal control of three-phase PWM inverter for UPS systems, in *IEEE 35th Annual Power Electronics Specialists Conference*, vol. *3*, 2004, pp. 2054–2059.

J.-W. Jung and A. Keyhani, Modeling and control of fuel cell based distributed generation systems in a standalone AC power supply, *J. Iranian Assoc. Electr. Electron. Eng.*, vol. *2*, no. 1, pp. 10–23, spring and summer 2005.

M. N. Marwali and A. Keyhani, Control of distributed generation systems, part I: voltage and Current Control, *IEEE Trans. Power Electron.*, vol. *19*, no. 6, pp. 1541–1550, Nov. 2004.

M. N. Marwali, J. W. Jung, and A. Keyhani, Control of distributed generation systems, part II: load sharing, *IEEE Trans. Power Electron.*, vol. *19*, no. 6, pp. 1551–1561, Nov. 2004.

J.-W. Jung and A. Keyhani, Design of Z-source converter for fuel cells, Electric Supply Industry in Transition, AIT Thailand, 2004.

M. Dai, M. N. Marwali, J.-W. Jung, and A. Keyhani, Power flow control of a single distributed generation unit with nonlinear local load, in *IEEE Power Engineering Society 2004 Power Systems Conference and Exposition*, vol. *1*, pp. 398–403, Oct. 2004.

P. T. Krein, R. S. Balog, and X. Geng, High-frequency link inverter for fuel cells based on multiple-carrier PWM, *IEEE Trans. Power Electron.*, vol. *19*, no. 5, pp. 1279–1288, Sept. 2004.

M. Orabi, T. Ninomiya, Y. Imai, and H. Nakagome, Review of pre-regulator CCM boost PFC converter dynamics limits, *IEEE 35th Annual Power Electronics Specialists Conference*, vol. *3*, 2004, pp. 2314–2319.

R. Teodorescu and F. Blaabjerg, Flexible control of small wind turbines with grid failure detection operating in stand-alone and grid-connected mode, *IEEE Trans. Power Electron.*, vol. *19*, no. 5, 2004, pp. 1323–1332.

The Mathworks, Control System Toolbox User's Guide Version 6, The Mathworks Inc., Natick, MA, 2004.

T. S. Lee, K. S. Tzeng, and M. S. Chong, Robust controller design for a single-phase UPS inverter using u-synthesis, *IEEE Proc. Electric Power Appl.*, vol. *151*, no. 3, pp. 334–340, 2004.

M. Orabi and T. Ninomiya, Study of alternative regimes to analyze two-stage PFC converter, in *Nineteenth Annual IEEE Applied Power Electronics Conference and Exposition*, vol. *3*, 2004, pp. 1488–1494.

D. Fournier and E. Westervelt, A candidate army energy and water management strategy, U.S. Army Engineer Research and Development Center, Champaign, IL, 2004.

K. Schoder and A. Feliachi, Simulation of electric shipboard power systems using Modelica and Matlab/Simulink, in *IEEE Power Engineering Society Meeting*, Denver, Colorado, June 2004.

K. Sedghisigarchi and A. Feliachi, Dynamic and transient analysis of power distribution systems with fuel cells—part I: Fuel-cell dynamic model, *IEEE Trans. Energy Conversion*, vol. *19*, no. 2, pp. 423–428, June 2004.

K. Sedghisigarchi and A. Feliachi, Dynamic and transient analysis of power distribution systems with fuel cells—part II: control and stability enhancement, *IEEE Trans. Energy Conversion*, vol. *19*, issue 2, pp. 429–434, June 2004.

R. H. Lasseter and P. Piagi, Microgrid: a conceptual solution, *PESC '04*, Aachen, Germany 20–25 June 2004.

G. Dong and O. Ojo, A generalized over-modulation methodology for current regulated three-phase voltage source converters, in *Proceedings 39th Annual Meeting of the IEEE Industry Applications (IAS '04)*, vol. *4*, Seattle, WA, Oct. 2004, pp. 2216–2223.

G. Dong and O. Ojo, An adaptive controller in stationary reference frame for d-statcom in unbalanced operation, *IEEE Trans. Ind. Electron.*, vol. *51*, no. 2, pp. 401–409, Apr. 2004.

Z. Y. Hou and J. Sun, Study on control strategy for three-phase four-leg inverter power supply, in *Proceedings 30th Annual Conference of the IEEE Industrial Electronics Society (IECON '04)*, vol. *1*, Busan, Korea, Nov. 2004, pp. 805–809.

T. S. Lee, K. S. Tzeng, and M. S. Chong, Robust controller design for a single-phase UPS inverter using μ-synthesis, *IEE Proc. Electr. Power Appl.*, vol. *151*, no. 3, pp. 334–340, Sept. 2004.

Y. W. Li, D. M. Vilathgamuwa, and P. C. Loh, Micro-grid power quality enhancement using a three-phase four-wire grid-interfacing compensator, in *Proceedings 39th Annual Meeting of the IEEE Industry Applications (IAS '04)*, vol. *3*, Seattle, WA, Oct. 2004, pp. 1439–1446.

H. Ma and Z. Cai, Comparison of control strategies for active power filter in three-phase four-wire systems, in *Proceedings 30th Annual Conference of the IEEE Industrial Electronics Society (IECON '04)*, vol. *2*, Busan, Korea, Nov. 2004, pp. 1429–1434.

K. Sundareswaran and A. P. Kumar, Voltage harmonic elimination in PWM A. C. chopper using genetic algorithm, *IEE Proc. Electr. Power Appl.*, vol. *151*, no. 1, pp. 26–31, Jan. 2004.

R. Teodorescu and F. Blaabjerg, Flexible control of small wind turbines with grid failure detection operating in stand-alone and grid-connected mode, *IEEE Trans. Power Electron.*, vol. *19*, no. 5, pp. 1323–1332, Sept. 2004.

X. Wei, J. H. Chow, B. Fardanesh, and A.-A. Edris, A common modeling framework of voltage-sourced converters for load flow, sensitivity, and dispatch analysis, *IEEE Trans. Power Syst.*, vol. *19*, no. 2, pp. 934–941, May 2004.

Z. M. Ye, P. K. Jain, and P. C. Sen, Robust controller design for high frequency resonant inverter system with voltage mode control, in *Proceedings 30th Annual Conference of the IEEE Industrial Electronics Society (IECON '04)*, vol. *1*, Busan, Korea, Nov. 2004, pp. 41–46.

K. Schoder, A. Hasanovic, A. Feliachi, and A. Hasanovic, PAT: A Power Analysis Toolbox for MATLAB/Simulink, *IEEE Trans. Power Syst.*, vol. *18*, no. 1, pp. 42–47, Feb. 2003.

K. Kwasinski, P. T. Krein and P. L. Chapman, Time domain comparison of pulse-width modulation schemes, *IEEE Power Electron Lett.*, vol. *1*, No. 3, pp. 64–68, September. 2003.

M. Dai, A. Keyhani, J. W. Jung, and. A. B., Proca, A low cost fuel cell drive system for electrical vehicles, in *Global Power train Congress '03 Conference and Exposition on Advanced Propulsion Strategy*, 2003, pp. 22–26.

O. J. Joyner and T. Abdallah, Miscellaneous Publication, ERDC/CERL, Champ aign, IL, Report Number ERDC/CERL LR-03-5. http://www.cecer.army.mil/td/tips/pub, Aug 19, 2003.

Consortium for Electric Reliability Technology Solutions (CERTS), Integration of Distributed Energy Resources: The CERTS MicroGrid Concept, Consultant report P500-03, California Energy Commission, http://certs.lbl.gov/pdf/50829.pdf, 2003.

P. Krishnamurthy, F. Khorrami, and R. S. Chandra, Global high-gain based observer and back stepping controller for generalized output-feedback canonical form, *IEEE Trans. Autom. Control*, vol. *48*, no. 12, pp. 2277–2283, Dec. 2003.

A. Keyhani, Leader-follower framework for control of energy services, *IEEE Trans. Power Syst.*, vol. *18*, issue 2, pp. 837–841, 2003.

F. Z. Peng, Z-source inverter, *IEEE Trans. Ind. Appl.*, vol. *39*, no. 2, pp. 504–510, 2003.

A. B. Proca, A. Keyhani, and J. M. Miller, Sensorless sliding-mode control of induction motors using operating condition dependent models, *IEEE Trans. Energy Conversion*, vol. *18*, no. 2, pp. 205–212, 2003.

R. Gemmen and P. Famouri, Electrochemical circuit model of a PEM fuel cell, in *IEEE Power Engineering Society 2003 General Meeting*, Toronto, Canada, 2003.

A. Chowdhury, S. K. Agarwal, and D. O. Koval, Reliability modeling of distributed generation in conventional distribution systems planning and analysis, *IEEE Trans. Ind. Appli.*, vol. *39*, pp. 1493–1498, 2003.

H. Mori, K. Matsui, K. Kondo, I. Yamamoto, and M. Hasegawa, Parallel-connected five-level PWM inverters, *IEEE Trans. Power Electron.*, vol. *18*, no. 1, pp. 173–79, 2003.

H. B. Puttgen, P. R. MacGrego, and F. C. Lambert, Distributed generation: semantic hype or the dawn of a new era? *IEEE Power Energ. Mag.*, vol. *1*, no. 1, pp. 22–29, 2003.

L. Y. Chiu and B. M. Diong, An improved small-signal model of the dynamic behavior of PEM fuel cells, in *IEEE 38th IAS Annual Meeting*, vol. *2*, pp. 709–715, 2003.

M. Chomat and L. Schreier, Control method for DC-link voltage ripple cancellation in voltage source inverter under unbalanced three-phase voltage supply conditions, in *Proceedings IEEE International Electric Machines and Drive Conference (IEMDC '03)*, vol. *2*, Madison, WI, June 2003, pp. 869–875.

W. El-Khattam, Y. G. Hegazy, and M. M. A. Salama, Stochastic power flow analysis of electrical distributed generation systems, in *Proceedings 2003 IEEE Power Engineering Society General Meeting*, vol. *2*, Toronto, Ontario, Canada, July 2003, pp. 1141–1144.

A. Kwasinski, P. T. Krein, and P. L. Chapman, Time domain comparison of pulse-width modulation schemes, *IEEE Power Electron. Lett.*, vol. *1*, no. 3, pp. 64–68, Sept. 2003.

S.-H. Lee and C.-C. Chu, Power flow models of unified power flow controllers in various operation modes, in *Proceedings 2003 IEEE PES Transmission and Distribution Conference and Exposition*, vol. *1*, Dallas, TX, Sept. 2003, pp. 157–162.

X. Li, P. Zhu, Y. Yang, and J. Chen, Experimental investigation of a three-phase series-parallel compensated UPS for non-linear load, in *Proceedings 29th Annual Conference of the IEEE Industrial Electronics Society (IECON '03)*, vol. *1*, Roanoke, VA, Nov. 2003, pp. 794–799.

X, Li, P. Zhu, Y. Yang, and J. Chen, Hybrid control of multiple inverters in an island-mode distribution system, in Proceedings. *2003 IEEE Power Electronics Specialists Conference (PESC '03)*, vol. *1*, Acapulco, Mexico, June 2003, pp. 61–66.

X. Li, P. Zhu, Y. Yang, and J. Chen, Three-phase unity-power-factor star-connected switch (VIENNA) rectifier with unified constant-frequency integration control, *IEEE Trans. Power Electron.*, vol. *18*, no. 4, pp. 952–957, July 2003.

R. M. Tallam, R. Naik, M. L. Gasperi, T. A. Nondahl, H. Lu, and Q. Yin, Practical issues in the design of active rectifiers for AC drives with reduced DC-link capacitance, in *Proceedings. 38th Annual Meeting of the IEEE Industry Applications (IAS '03)*, vol. *3*, Salt Lake City, UT, Oct. 2003, pp. 1538–1545.

M. T. Tsai and C. H. Liu, Design and implementation of a cost-effective quasi line-interactive UPS with novel topology, *IEEE Trans. Power Electron.*, vol. *18*, no. 4, pp. 1002–1011, July 2003.

G. E. Valderrama, A. M. Stankovic, and P. Mattavelli, Dissipativity-based adaptive and robust control of UPS in unbalanced operation, *IEEE Trans. Power Electron.*, vol. *18*, no. 4, pp. 1056–1062, July 2003.

K. Zhang, Y. Kang, J. Xiong, and J. Chen, Direct repetitive control of SPWM inverter for UPS purpose, *IEEE Trans. Power Electron.*, vol. *18*, no. 3, pp. 784–792, May 2003.

D. N. Zmood and D. G. Holmes, Stationary frame current regulation of PWM inverters with zero steady-state error, *IEEE Trans. Power Electron.*, vol. *18*, no. 3, pp. 814–822, May 2003.

S. Yerramalla, A. Davari, and A. Feliachi, Dynamic modeling and analysis opolymer electrolyte fuel cell, *IEEE Power Engineering Society Summer Meeting*, vol. *1*, pp. 82–86, 2002.

J. T. Pukrushpan, A. G. Stefanopoulou, and H. Peng, Modeling and control for PEM fuel cell stack system, *Am. Control Conf.*, vol. *4*, pp. 3117–3122, 2002.

J. Hatziadoniu, A. A. Lobo, F. Pourboghrat, and M. Daneshdoost, A simplified dynamic model of grid-connected fuel-cell generators, *IEEE Trans. Power Delivery*, vol. *17*, pp. 467–473, 2002.

H. Nikkhajoei, and M. R. Iravani, Modeling and analysis of a micro-turbine generation system in *IEEE Power Engineering Society Summer Meeting*, vol. *1*,2002, pp. 167–169.

D. Niebur, R. Stoicescu, Xiaoguang Yang, 3-phase converter modeling for unbalanced radial systems, K. Miu, and C. O. Nwankpa, in *Transmission and Distribution Conference and Exhibition 2002, Asia Pacific, IEEE/PES*, vol. *1*, 6–10 Oct. 2002, pp. 563–567.

R. Stoicescu, K. Miu, C. O. Nwankpa, D. Niebur, and X. Yang, Three-phase converter models for unbalanced radial power-flow studies, *IEEE Trans. Power Syst.*, vol. *17*, issue 4, pp. 1016–1021, Nov. 2002.

R. Datta and V. T. Ranganathan, Variable-speed wind power generation using doubly-fed wound rotor induction machine — a comparison with alternative schemes, *IEEE Trans. Energy Conversion*, vol. *17*, no. 3, pp. 414–420, Sept. 2002.

J. T. Pukrushpan, H. Peng, and A. G. Stefanopoulou, Simulation and analysis of transient fuel cell system performance based on a dynamic reactant flow model, *Proceedings of ASME IMECE '02*, 2002.

P. Krishnamurthy and F. Khorrami, Generalized adaptive output-feedback form with unknown parameters multiplying high output relative-degree states, in *41st IEEE Conference on Decision and Control*, vol. *2*, 2002, pp. 1503–1508.

P. Krishnamurthy, F. Khorrami, and Z. P. Jiang, Global output feedback tracking for nonlinear systems in generalized output-feedback canonical form, *IEEE Trans. Autom. Control*, vol. *47*, no. 5, pp. 814–19, 2002.

X. Lin, X. Chen, Y. Kang, S. Duan, and J. Chen, Parallel three-phase UPS inverters with a new control technique, in *33rd Annual IEEE Power Electronics Specialists Conference*, vol. *2*, 2002, pp. 905–908.

T. Uematsu, K. Tanaka, Y. Takayanagi, H. Kawasaki, and T. Ninomiya, Utility interactive inverter controllable for a wide range of DC input voltage, in *Proceedings of the IEEE Power Conversion Conference*, vol. *2*, 2002, pp. 498–503.

K. Siri and K. A. Conner, Parallel-connected converters with maximum power tracking, in *17th Annual IEEE Applied Power Electronics Conference and Exposition*, vol. *1*, 2002, pp. 419–425.

M. Shahidehpour, Distributed resources for transmission congestion and expansion management, in *IEEE Power Engineering Society Summer Meeting*, vol. *3*, 2002, pp. 1045–1048.

C. J. Hatziadoniu, A. A. Lobo, F. Pourboghrat, and M. Daneshdoost, A simplified dynamic model of grid-connected fuel-cell generators, *IEEE Trans. Power Delivery*, vol. *17*, no. 2, pp. 467–473, 2002.

K. Sedghisigarchi and A. Feliachi, Control of grid-connected fuel cell power plant for transient stability enhancement, in *IEEE Power Engineering Society Winter Meeting*, vol. *1*, 2002, pp. 383–388.

R. Gopinath, S. Kim, J.-H. Hahn, M. Webster, J. Burghardt, S. Campbell, D. Becker, P. Enjeti, M. Yeary, and J. Howze, Development of a low cost fuel cell inverter system with DSP control, *IEEE PESC '02*, vol. *1*, pp. 309–314, 2002.

Y. Zhu and K. Tomsovic, Adaptive power flow method for distribution systems with dispersed generation, *IEEE Trans. Power Delivery*, vol. *17*, pp. 822–827, 2002.

R. Lasseter, A. Akhil, C. Marnay, J. Stephens, J. Dagle, R. Guttromson, A. S. Meliopoulous, R. Yinger, and J. Eto, Integration of distributed energy resources: the CERTS MicroGrid concept, available from: http://certs.lbl.gov, April 2002.

S. Jemei, D. Hissel, M. C. Pera, and J. M. Kauffmann, Black-box modeling of proton exchange membrane fuel cell generators, pp. 1474–1478, 2002.http://certs.lbl.gov, April 2002.

S. A. O. da Silva, P. F. Donoso-Garcia, P. C. Cortizo, and P. F. Seixas, A three-phase line-interactive UPS system implementation with series–parallel active power-line conditioning capabilities, *IEEE Trans. Ind. Appl.*, vol. *38*, no. 6, pp. 1581–1590, Nov./Dec. 2002.

M. Illindala and G. Venkataramanan, Control of distributed generation systems to mitigate load and line imbalances, in *Proceedings 2002 IEEE Power Electronics Specialists Conference (PESC '02)*, vol. *4*, Cairns, Queensland, Australia, June 2002, pp. 2013–2018.

A. Keyhani, M. Dai, and M. N. Marwali, UPS overload protection and current control using discrete-time sliding mode technique, in *Proceedings IASTED International Conference on Power and Energy Systems*, Marina del Rey, CA, USA, May 2002, pp. 504–510.

T. K. Lee, W. C. Lee, H. Jang, and D. S. Hyun, A study on the line-interactive ups using the series voltage compensator, in *Proceedings 28th Annual Conference of the IEEE Industrial Electronics Society (IECON '02)*, vol. *1*, Sevilla, Spain, Nov. 2002, pp. 733–738.

J. Liang, T. C. Green, G. Weiss, and Q. C. Zhong, Evaluation of repetitive control for power quality improvement of distributed generation, in *Proceedings 2002 IEEE Power Electronics Specialists Conference (PESC '02)*, vol. *4*, Cairns, Queensland, Australia, June 2002, pp. 1803–1808.

N. Mendalek, F. Fnaiech, K. Al-Haddad, and L. A. Dessaint, A non-linear optimal predictive control of a shunt active power filter, in *Proceedings 37th Annual Meeting of the IEEE Industry Applications (IAS '02)*, vol. *1*, Pittsburgh, PA, Oct. 2002, pp. 70–77.

L. Mihalache, DSP control method of single-phase inverters for ups applications, in *Proceedings IEEE Applied Power Electronics Conference and Exposition (APEC '02)*, vol. *1*, Dallas, TX, Mar. 2002, pp. 590–596.

T. L. Tai and J. S. Chen, UPS inverter design using discrete-time sliding mode control scheme, *IEEE Trans. Ind. Electron.*, vol. *49*, no. 1, pp. 67–75, Feb. 2002.

R. Zhang, V. H. Prasad, D. Boroyevich, and F. C. Lee, Three-dimensional space vector modulation for four-leg voltage-source converters, *IEEE Trans. Power Electron.*, vol. *17*, no. 3, pp. 314–326, May 2002.

K. Zhou and D. Wang, Relationship between space-vector modulation and three-phase carrier-based PWM: a comprehensive analysis, *IEEE Trans. Ind. Electron.*, vol. *49*, no. 1, pp. 186–196, Feb. 2002.

Eduard Muljadi and C. P. Butterfield, Pitch-Controlled Variable-speed Wind turbine Generation, *IEEE Trans. Ind. Appl.*, vol. *37*, no. 1, pp. 240–246, Jan./Feb. 2001.

Y. K. Fan, D. Niebur, C. O. Nwankpa, H. Kwatny, R. Fischl, Voltage dynamics of small integrated AC/DC power systems, in *Proceedings of American Control Conference*, vol. 2, June 25–27, 2001, pp. 829–830.

J. L. Del Monaco, The role of distributed generation in the critical electric power infrastructure, in *IEEE-Power Engineering Society Winter Meeting*, vol. *1*, 2001, pp. 144–145.

M. D. Lukas, K. Y. Lee, and H. Ghezel-Ayagh, An explicit dynamic model for direct reforming carbonate fuel cell stack, *IEEE Trans. Energy Conversion*, vol. *16*, pp. 289–295, 2001.

U. Borup, F. Blaabjerg, and P. N. Enjeti, Sharing of nonlinear load in parallel-connected three-phase converters, *IEEE Trans. Ind. Appli.*, vol. *37*, no. 6, pp. 1817–1823, 2001.

Qiao and K. M. Smedley, Three-phase grid-connected inverters interface for alternative energy sources with unified constant-frequency integration control, *Conference Record of the 2001 IEEE Industry Applications Conference*, vol. *4*, 2001, pp. 2675–2682.

B. H. Kwon, J. H. Choi, and T. W. Kim, Improved single-phase line-interactive UPS, *IEEE Trans. Ind. Electron.*, vol. *48*, no. 4, 2001, pp. 804–811.

J. A. McDowall, Opportunities for electricity storage in distributed generation and renewables, in *IEEE/PES Transmission and Distribution Conference and Exposition*, vol. 2, 2001, pp. 1165–1168.

R. Lasseter, Dynamic models for micro-turbines and fuel cells, in *IEEE Power Engineering Society Summer Meeting*, Vancouver, BC, Canada, vol. 2, 2001, pp. 761–766.

S. Rahman, Fuel cell as a distributed generation technology, in *IEEE Power Engineering Society Summer Meeting*, Vancouver, BC, Canada, vol. *1*, 2001, p. 551–552.

M. C. Williams, Fuel cells and the world energy future, in *IEEE Power Engineering Society Summer Meeting*, Vancouver, BC, Canada, vol. *1*, 2001, p. 725.

Price and B. J. Davidson, Recent developments in the design and applications of utility-scale energy storage plant, in *16th IEEE International Conference and Exhibition on Electricity Distribution*, Amsterdam, Netherlands, 2001, p. 4.

S. Gilbert, The nations largest fuel cell project, a 1 MW fuel cell power plant deployed as a distributed generation resource, Anchorage, Alaska project dedication August 9, 2000, *IEEE Rural Electric Power Conference*, Little Rock, AR, 2001, pp. A4/1–8.

B. Lasseter, Microgrids [distributed power generation], *IEEE Power Engineering Society Winter Meeting*, Columbus, OH, vol. *1*, pp. 146–149, 2001.

J. L. Del Monaco, The role of distributed generation in the critical electric power infrastructure, in *IEEE Power Engineering Society Winter Meeting*, Columbus, OH, vol. *1*, 2001, pp. 144–145.

C. Marnay, F. J. Rubio, and A. S. Siddiqui, Shape of the microgrid, in *IEEE Power Engineering Society Winter Meeting*, Columbus, OH, vol. *1*, 2001, pp. 150–153.

M. Etezadi-Amoli and K. Choma, Electrical performance characteristics of a new microturbine generator, *IEEE Power Engineering Society Winter Meeting*, Columbus, OH, vol. *2*, 2001, pp. 736–740.

K. Tomsovic and T. Hiyama, Intelligent control methods for system with dispersed generation, in *IEEE Power Engineering Society Winter Meeting*, Columbus, OH, vol. *2*, 2001, pp. 913–917.

A. Bonhomme, D. Cortinas, F. Boulanger, and J. L. Fraisse, A new voltage control system to facilitate the connection of dispersed generation to distribution networks, in *CIRED IEEE-16th International Conference and Exhibition on Electricity Distribution*, vol. *4*, no. 482, 2001, p. 248.

Serrani, A. Isidori, and L. Marconi, Semi-global nonlinear output regulation with adaptive internal model, *IEEE Trans. Autom. Control*, vol. *46*, no. 8, pp. 1178–1940, 2001.

T. Hu and Z. Lin, *Control Systems with Actuator Saturation: Analysis and Design*, Birkhauser, Boston, 2001.

S. J. Chiang, C. Y. Yen, and K. T. Chang, A multimodule parallelable series-connected PWM voltage regulator, *IEEE Trans. Ind. Electron.*, vol. *48*, no. 3, pp. 506–516, 2001.

J. L. Del Monaco, The role of distributed generation in the critical electric power infrastructure, in *IEEE Power Engineering Society Winter Meeting*, vol. *1*, 2001, pp. 144–145.

Giraud, F. Salameh Z.M., Steady-state performance of a grid-connected rooftop hybrid wind photovoltaic power system with battery storage, *IEEE Trans. Energy Conversion*, vol. *16*, no. 1, pp. 1–7, 2001.

U. Borup, P. N. Enjeti, and F. Blaabjerg, A new space-vector-based control method for UPS systems powering nonlinear and unbalanced loads, *IEEE Trans. Ind. Appl.*, vol. *37*, no. 6, pp. 1864–1870, Nov./Dec. 2001.

F. Botteron, H. Pinheiro, H. A. Grundling, and J. R. P. H. L. Hey, Digital voltage and current controllers for three-phase PWM inverter for UPS applications, in *Proceedings. 36th Annual Meeting of the IEEE Industry Applications (IAS '01)*, vol. *4*, Chicago, IL, Sept./Oct. 2001, pp. 2667–2674.

S. Buso, S. Fasolo, and P. Mattavelli, Uninterruptible power supply multi-loop control employing digital predictive voltage and current regulators, *IEEE Trans. Ind. Appl.*, vol. *37*, no. 6, pp. 1846–1854, Nov./Dec. 2001.

H. Dehbonei, C. Nayar, L. Borle, and M. Malengret, A solar photovoltaic in-line UPS system using space vector modulation technique, in *Proceedings. 2001 IEEE Power Engineering Society Summer Meeting*, vol. *1*, Vancouver, BC, Canada, July 2001, pp. 632–637.

S. El-Barbari and W. Hofmann, Control of a 4 leg inverter for standalone photovoltaic systems, in *Proceedings. 4th IEEE International Conference on Power Electronics and Drive Systems (PEDS '01)*, vol. *1*, Denpasar, Indonesia, Oct. 2001, pp. 348–354.

G. Escobar, A. M. Stankovic, and P. Mattavelli, Dissipativity-based adaptive and robust control of UPS in unbalanced operation, in *Proceedings. 2001 IEEE Power Electronics*

Specialists Conference (PESC '01), vol. *4*, Vancouver, BC, Canada, June 2001, pp. 1862–1867.

S. Fukuda and T. Yoda, A novel current-tracking method for active filters based on a sinusoidal internal model, *IEEE Trans. Ind. Appl.*, vol. *37*, no. 3, pp. 888–895, May/June 2001.

W. Guo, S. Duan, Y. Kang, and J. Chen, A new digital multiple feedback control strategy for single-phase voltage-source PWM inverters, in *Proceedings IEEE Region 10 International Conference on Electrical and Electronic Technology (TENCON '01)*, vol. 2, Singapore, Aug. 2001, pp. 809–813.

W. Guo, S. Duan, X. Kong, Y. Kang, and J. Chen, A modified deadbeat control for single-phase voltage-source PWM inverters based on asymmetric regular sample, in *Proceedings IEEE Applied Power Electronics Conference and Exposition(APEC '01)*, vol. 2, Vancouver, BC, Canada, June 2001, pp. 962–967.

B. H. Kwon, J. H. Choi, and T. W. Kim, Improved single-phase line-interactive UPS, *IEEE Trans. Ind. Electron.*, vol. *48*, no. 4, pp. 804–811, Aug. 2001.

B. Lasseter, Microgrids, in *Proceedings 2001 IEEE Power Engineering Society Winter Meeting*, vol. *1*, Columbus, OH, Jan.–Feb. 2001, pp. 146–149.

T. S. Lee, S. J. Chiang, and J. M. Chang, $H\infty$ loop-shaping controller designs for the single-phase UPS inverters, *IEEE Trans. Power Electron.*, vol. *16*, no. 4, pp. 473–481, July 2001.

P. C. Loh, M. J. Newman, D. N. Zmood, and D. G. Holmes, Improved transient and steady state voltage regulation for single and three phase uninterruptible power supplies, in *Proceedings 2001 IEEE Power Electronics Specialists Conference(PESC '01)*, vol. *2*, Vancouver, BC, Canada, June 2001, pp. 498–503.

P. Mattavelli, Synchronous-frame harmonic control for high-performance AC power supplies, *IEEE Trans. Ind. Appl.*, vol. *37*, no. 3, pp. 864–872, May/June 2001.

P. Mattavelli, G. Escobar, and A. M. Stankovic, Dissipativity-based adaptive and robust control of UPS, *IEEE Trans. Ind. Electron.*, vol. *48*, no. 2, pp. 334–343, Apr. 2001.

J. L. D. Monaco, The role of distributed generation in the critical electric power infrastructure, in *Proceedings 2001 IEEE Power Engineering Society Winter Meeting*, vol. *1*, Columbus, OH, Jan.–Feb. 2001, pp. 144–145.

D. Noriega-Pineda, G. Espinosa-Perez, A. Varela-Vega, and S. Horta-Mejia, Experimental evaluation of an adaptive nonlinear controller for single-phase UPS, in *Proceedings 2001 IEEE International Conference on Control Applications(CCA '01)*, Mexico City, Mexico, Sept. 2001, pp. 254–258.

S. Ponnaluri, A. Brickwedde, and R. W. D. Doncker, Overriding individual harmonic current control with fast dynamics for UPS with non-linear loads, in *Proceedings 4th IEEE International Conference on Power Electronics and Drive Systems (PEDS '01)*, vol. 2, Denpasar, Indonesia, Oct. 2001, pp. 527–532.

C. Qiao and K. M. Smedley, Three-phase grid-connected inverters interface for alternative energy sources with unified constant-frequency integration control, in *Proceedings 2001 IEEE Industry Applications Society 36th Annual Meeting (IAS '01)*, vol. *4*, Chicago, IL, Sept.–Oct. 2001, pp. 2675–2682.

C. Rech, H. Pinheiro, H. A. Gräundling, H. L. Hey, and J. R. Pinheiro, Analysis and design of a repetitive predictive-PID controller for PWM inverters, in *Proceedings IEEE Applied Power Electronics Conference and Exposition (APEC '01)*, vol. 2, Vancouver, BC, Canada, June 2001, pp. 986–991.

D. J. Smith, Distributed generation: The power to choose, *Power Eng*, vol. *105*, no. 3, pp. 32–35, Mar. 2001.

A. V. Stankovic and T. A. Lipo, A novel control method for input output harmonic elimination of the PWM boost type rectifier under unbalanced operating conditions, *IEEE Trans. Power Electron.*, vol. *16*, no. 5, pp. 603–611, Sept. 2001.

T. Takeshita, T. Masuda, and N. Matsui, Current waveform control of distributed generation system for harmonic voltage suppression, in *Proceedings 2001 IEEE Power Electronics Specialists Conference (PESC '01)*, vol. *2*, Vancouver, BC, Canada, June 2001, pp. 516–521.

S. R. Wall, Performance of inverter interfaced distributed generation, in *Proceedings 2001 IEEE/PES Transmission and Distribution Conference and Exposition*, vol. *2*, Atlanta, GA, Oct.–Nov. 2001, pp. 945–950.

D. N. Zmood, D. G. Holmes, and G. H. Bode, Frequency-domain analysis of three-phase linear current regulators, *IEEE Trans. Ind. Appl.*, vol. *37*, no. 2, pp. 601–610, Mar./Apr. 2001.

D. J. Tooth, , S. J. Finney, , B. W. Williams, , Effects of using DC-side average current-mode control on a three-phase converter with an input filter and distorted supply, *IEEE Proc. on Electric Power Appl.*, vol. *147*, no. 6, pp. 459–477, 2000.

H. M. Nehrir, J. B. Lameres, G. Venkataramanan, V. Gerez, and A. L. Alvarado, An approach to evaluate the general performance of stand-alone wind/photovoltaic generating systems, *IEEE Trans. Energy Conversion*, vol. *15*, issue 4, pp. 433–439, Dec. 2000.

Y. B. Byun, T. G. Koo, K. Y. Joe, E. S. Kim, J. I. Seo, and D. H. Kim, Parallel operation of three-phase UPS inverters by wireless load sharing control, *IEEE INTELEC*, Phoenix, AZ, pp. 526–532, 2000.

K. H. Hauer, Dynamic interaction between the electric drive train and fuel cell system for the case of an indirect methanol fuel cell vehicle, *35th IECEC Meeting*, vol. *2*, pp. 1317–1325, 2000.

N. Hur and K. Nam, A robust load-sharing control scheme for parallel-connected multi systems, *IEEE Trans. Ind. Electron.*, vol. *47*, pp. 871–879, 2000.

B. Maurhoff and G. Wood, Dispersed generation to reduce power costs and improve service reliability, *IEEE Rural Electric Power Conference*, Louisville, KY, pp. C5/1–C5/7, 2000.

L. Philipson, Distributed and dispersed generation: addressing the spectrum of consumer needs, *IEEE Power Engineering Society Summer Meeting*, Seattle, WA, vol. *3*, pp. 1663–1665, 2000.

R. K. Jardan, I. Nagy, T. Nitta, and H. Ohsaki, Power factor correction in a turbine-generator-converter system, *IEEE-IAS Conf. Record*, Rome, Italy, vol. *2*, pp. 894–900, 2000.

L.J. J. Offringa and J. L. Duarte, A 1600 kW IGBT converter with interphase transformer for high speed gas turbine power plants, *IEEE-IAS Conf. Record*, Rome, Italy, vol. *4*, pp. 2243–2248, 2000.

U. B. Jensen, F. Blaabjerg, and P. N. Enjeti, Sharing of nonlinear load in parallel connected three-phase converters, *IEEE-IAS Conf. Record*, Rome, Italy, vol. *4*, pp. 2338–2344, 2000.

Tuladhar, H. Jin, T. Unger, and K. Mauch, Control of parallel inverters in distributed AC power systems with consideration of line impedance effect, *IEEE Trans. Ind. Appl.*, vol. *36*, no. 1, pp. 131–38, 2000.

F. Yoshiba, T. Abe, and T. Watanabe, Numerical analysis of molten carbonate fuel cell stack performance: diagnosis of internal conditions using cell voltage profiles, *J. Power Sources*, vol. *87*, no. 1–2, pp. 21–27, 2000.

IEEE Recommended Practice for Utility Interface of Photovoltaic (PV) Systems, IEEE Standard 929–2000.

S. Chen and G. Joos, Transient performance of UPS system with synchronous-frame digital controller, in *Proceedings 22nd International Telecommunications Energy Conference (INTELEC '00)*, Phoenix, AZ, Sept. 2000, pp. 533–540.

P.-T. Cheng, S. Bhattacharya, and D. Divan, Experimental verification of dominant harmonic active filter for high-power applications, *IEEE Trans.* Ind. *Appli.*, vol. *36*, no. 2, pp. 567–577, Mar./Apr. 2000.

P.-T. Cheng, S. Bhattacharya, and D. Divan, A comparative analysis of control algorithms for three-phase line-interactive UPS systems with series-parallel active power-line conditioning using srf method, in *Proceedings IEEE Applied Power Electronics Conference and Exposition (APEC '00)*, vol. 2, Galway, Ireland, June 2000, pp. 1023–1028.

Y. K. Fan, D. Niebur, C. O. Nwankpa, H. Kwatny, and R. Fischl, Saddle-node bifurcation of voltage profiles of small integrated AC/DC power systems, in *Proceedings 2000 IEEE Power Engineering Society Summer Meeting*, vol. *1*, Seattle,WA, USA, July 2000, pp. 614–619.

T. D. Knapp and H. M. Budman, Robust control design of nonlinear processes using empirical state affine models, *Int. J. Control*, vol. *73*, no. 17, pp. 1525–1535, Nov. 2000.

D. C. Lee, A common mode voltage reduction in boost rectifier/inverter system by shifting active voltage vector in a control period, *IEEE Trans. Power Electron.*, vol. *15*, no. 6, pp. 1094–1101, Nov. 2000.

J. C. Liao and S. N. Yeh, A novel instantaneous power control strategy and analytic model for integrated rectifier/inverter systems, *IEEE Trans. Power Electron.*, vol. *15*, no. 6, pp. 996–1006, Nov. 2000.

W.-L. Lu, S.-N. Yeh, J.-C. Hwang, and H.-P. Hsieh, Development of a single phase half-bridge active power filter with the function of uninterruptible power supplies, *IEE Proc. Electr. Power Appl.*, vol. *147*, no. 4, pp. 313–319, July 2000.

V. F. Montagner, E. G. Carati, and H. A. Gräundling, An adaptive linear quadratic regulator with repetitive controller applied to uninterruptible power supplies, in *Proceedings 35th Annual Meeting of the IEEE Industry Applications (IAS '00)*, vol. *4*, Rome, Italy, Oct. 2000, pp. 2231–2236.

C. Rech, H. A. Gräundling, and J. R. Pinheiro, Comparison of discrete control techniques for UPS applications, in *Proceedings 35th Annual Meeting of the IEEE Industry Applications (IAS '00)*, vol. *4*, Rome, Italy, Oct. 2000, pp. 2531–2537.

V. Utkin, J. Guldner, and J. Shi, *Sliding Mode Control in Electromechanical Systems*, Taylor & Francis, Philadelphia, PA, 1999.

Y. Tzou, S. Jung, and H. Yeh, Adaptive repetitive control of PWM inverters for very low THD AC-voltage regulation with unknown loads, *IEEE Trans. Power Electron.*, vol. *14*, no. 5, pp. 973–981, 1999.

M. N. Marwali, and A. Keyhani, PWM control of three-phase series resonant DC link converters for UPS applications, *Thirty-Fourth IAS Annual Meeting. Conference Record of Industry Applications Conference*, vol. *3*, pp. 2017–2024, 1999.

S. J. Huang and J. C. Wu, A control algorithm for three-phase three-wired active power filters under nonideal mains voltages, *IEEE Trans. Power Electron.*, vol. *14*, pp. 753–760, 1999

E. Barsoukov, J. H. Kim, C. O. Yoon, and H. Lee, Universal battery parameterization to yield a non-linear equivalent circuit valid for battery simulation at arbitrary load, *J. Power Sources*, vol. *83*, no. 1, pp. 61–70, 1999.

W. Turner, M. Parten, D. Vines, J. Jones, and T. Maxwell, Modeling a PEM fuel cell for use in a hybrid electric vehicle, *Proceedings of the 1999 IEEE 49th Vehicular Technology Conference*, Houston, TX, vol. *2*, pp. 1385–1388, 1999.

J. Hall and R. G. Colclaser, Transient modeling and simulation of a tubular solid oxide fuel cell, *IEEE Trans. Energy Conversion*, vol. *14*, pp. 749–753, 1999.

M. D. Lukas, K. Y. Lee, and H. Ghezel-Ayagh, Development of a stack simulation model for control study on direct reforming molten carbonate fuel cell power plant, *IEEE Trans. Energy Conversion*, vol. *14*, pp. 1651–1657, 1999.

L. K. Wong, F. H. F. Leung, P. K. S. Tam, Control of PWM inverter using a discrete-time sliding mode controller, in *Proceedings of the IEEE 1999 Power Electronics and Drive Systems*, vol. *2*, 1999, pp. 947–950.

S. I. Moon, A. Keyhani, S. Pillutla, Nonlinear Neural-Network Modeling of an Induction Machine, *IEEE Trans. Control Sys. Technol.*, vol. *7*, no. 2, pp. 203–211, 1999.

J. S. Bay, *Fundamentals of Linear State Space Systems,*. WCB/McGraw-Hill, New York, 1999.

S. A. O. da Silva, P. F. Donoso-Garcia, and P. C. Cortizo, A three-phase series–parallel compensated line-interactive UPS system with sinusoidal input current and sinusoidal output voltage, in *Proceedings 34th Annual Meeting of the IEEE Industry Applications (IAS '99)*, vol. *2*, Phoenix, AZ, USA, Oct. 1999, pp. 826–832.

H. Äozbay, *Introduction to Feedback Control Theory*. New York, CRC Press, 1999.

J. Balcells, and D. Gonzalez, Harmonics due to resonance in a wind power plant, in *8th International Conference on Harmonics and Quality of Power*, vol. *2*,1998, pp. 896–899.

R. Chiang and M. Safonov, *Robust Control Toolbox*, The Mathworks, Natick, MA, 1998.

J. Cardell, M. Ilic, and R. Tabors, Integrating small scale distributed generation into a deregulated market: control strategies and price feedback, *Final Report to US Department of Energy*, MIT TR 98-001, April 1998.

H. A. Gräundling, E. G. Carati, and J. R. Pinheiro, Analysis and implementation of a modified robust model reference adaptive control with repetitive controller for UPS applications, in *Proceedings 24th Annual Conference of the IEEE Industrial Electronics Society (IECON '98)*, vol. *1*, Aachen, Germany, Aug.–Sept. 1998, pp. 391–395.

A. M. Mohamed, Modern robust control of a CSI-fed induction motor drive system, in *Proceedings 1998 American Control Conference*, vol. *6*, Philadelphia, PA, June 1998, pp. 3803–3808.

J. F. Moynihan, M. G. Egan, and J. M. D. Murphy, Theoretical spectra of space-vector-modulated waveforms, *IEE Proc. Electr. Power Appl.*, vol. *145*, no. 1, pp. 17–24, Jan. 1998.

Y. Sato, T. Ishizuka, K. Nezu, and T. Kataoka, A new control strategy for voltage-type PWM rectifiers to realize zero steady-state control error in input current, *IEEE Trans. Ind. Appl.*, vol. *34*, no. 3, pp. 480–486, May/June 1998.

V. Blasko, Analysis of a hybrid PWM based on modified space-vector and triangle-comparison methods, *IEEE Trans. Ind. Appl.*, vol. *33*, no. 3, pp. 756–764, May/June 1997.

S. R. Bowes and Y. S. Lai, The relationship between space-vector modulation and regular-sampled PWM, *IEEE Trans. Ind. Electron.*, vol. *44*, no. 5, pp. 670–679, Oct. 1997.

C. L. Chu, A three-phase four-wire unity power factor AC/DC converter with dual DC output voltage control, *Int. J. Electronics*, vol. *83*, no. 5, pp. 685–702, Nov. 1997.

W. Kolar and F. C. Zach, A novel three-phase utility interface minimizing line current harmonics of high-power telecommunications rectifier modules, *IEEE Trans. Ind. Electron.*, vol. *44*, no. 4, pp. 456–467, Aug. 1997.

Z. Lin, X. Bao, and B. M. Chen, Further results on almost disturbance decoupling with global asymptotic stability for nonlinear systems, in *Proceedings 36th Conference on Decision and Control*, vol. *3*, San Diego, CA, Dec. 1997, pp. 2847–2852.

H. Mao, F. C. Lee, D. Boroyevich, and S. Hiti, Review of high-performance three-phase power-factor correction circuits, *IEEE Trans. Ind. Electron.*, vol. *44*, no. 4, pp. 437–446, Aug. 1997.

C.-M. Ong, *Dynamic Simulation of Electric Machinery*. Prentice Hall, Upper Saddle River, NJ, 1997.

I. F. El-Sayed, Analysis, Design, and evaluation of a high performance PWM AC/DC converter providing sinusoidal current with unity power factor, *EPE J.*, vol. *6*, no. 3–4, pp. 36–45, 1996.

J. Cardell and M. Ilic, Modeling & stability analysis for distributed generation systems, in *Proceedings of North American Power Symposium*, MIT, Cambridge, MA, pp. 239–247, Nov. 1996.

M. Jankovic, P. Ninkovic, and Z. Jandra, A novel approach to current-mode control of the constant frequency power converters, in *Proceedings of the IEEE International Symposium on Industrial Electronics*, vol. *1*, pp. 488–492, 1996.

M. Jankovic, P. Ninkovic, and Z. Jandra, Analysis and design of a multiple feedback loop control strategy for single-phase voltage-source UPS inverters, *IEEE Trans. Power Electron.*, vol. *11*, no. 4, pp. 532–541, July 1996.

M. Jankovic, P. Ninkovic, and Z. Jandra, Small-signal model and analysis of a multiple feedback control scheme for three-phase voltage-source UPS inverters, in *Proceedings 1996 IEEE Power Electronics Specialists Conference (PESC '96)*, vol. *1*, Baveno, Italy, June 1996, pp. 188–194.

M. K. Donnelly, J. E. Dagle, D. J. Trudnowski, and G. J. Rogers, Impacts of the distributed utility on transmission system stability, *IEEE Trans. Power Syst.*, vol. *11*, no. 2, pp. 741–746, May 1996.

J. L. Lin and S. J. Chen, μ-based controller design for a DC–DC switching power converter with line and load variations, in *Proceedings 22th Annual Conference of the IEEE Industrial Electronics Society (IECON '96)*, vol. 2, Taipei, Taiwan, Aug. 1996, pp. 1029–1034.

S. Rathmann and H. A. Warner, New generation UPS technology, the delta conversion principle, in *Proceedings 31th Annual Meeting of the IEEE Industry Applications (IAS '96)*, vol. *4*, San Diego, CA, Oct. 1996, pp. 2389–2395.

P. Rioual, H. Pouliquen, and J. P. Louis, Regulation of a PWM rectifier in the unbalanced network state using a generalized model, *IEEE Trans. Power Electron.*, vol. *11*, no. 3, pp. 495–502, May 1996.

W. Shireen and M. S. Arefeen, An utility interactive power electronics interface for alternate/renewable energy systems, *IEEE Trans. Energy Conversion*, vol. *11*, no. 3, pp. 643–649, Sept. 1996.

Q. Yu, S. D. Round, L. E. Norum, and T. M. Undeland, Dynamic control of a unified power flow controller, in *Proceedings 1996 IEEE Power Electronics Specialists Conference (PESC '96)*, vol. *1*, Baveno, Italy, June 1996, pp. 508–514.

J. R. Espinoza, G. Joos, State variable decoupling and power flow control in PWM current source rectifiers, *21st International Conference on Industrial Electronics, Control, and Instrumentation*, vol. *1*, pp. 686–691, 1995.

D. Czarkowski, L. R. Pujara, and M. K. Kazimierczuk, Robust stability of state-feedback control of PWM DC-DC push–pull converter, *IEEE Trans. Ind. Electron.*, vol. *42*, no. 1, pp. 108–111, 1995.

J. Huang, Asymptotic tracking and disturbance rejection in uncertain nonlinear systems, *IEEE Trans. Autom. Control*, vol. *40*, no. 6, pp. 1118–1122, 1995.

R. K. Jardan, Generation of electric energy by high-speed turbine-generator sets, *IEEE/INTELEC '95*, The Hague, Netherlands, pp. 733–740, 1995.

D. Czarkowski, L. R. Pujara, and M. K. Kazimierczuk, Robust stability of state-feedback control of PWM DC–DC push–pull converter, *IEEE Trans. Ind.Electron.*, vol. *42*, no. 1, pp. 108–111, Feb. 1995.

Kamran and T. G. Habetler, An improved deadbeat rectifier regulator using a neural net predictor, *IEEE Trans. Power Electron.*, vol. *10*, no. 4, pp. 504–510, July 1995.

L. Malesani, L. Rossetto, P. Tenti, and P. Tomasin, AC/DC/AC PWM converter with reduced energy storage in the DC link, *IEEE Trans. Ind. Appl.*, vol. *31*, no. 2, pp. 287–292, Mar./Apr. 1995.

R. Naik, N. Mohan, M. Rogers, and A. Bulawka, A novel grid interface, optimized for utility-scale applications of photovoltaic, wind-electric, and fuel-cell systems, *IEEE Trans. Power Delivery*, vol. *10*, no. 4, pp. 1920–1926, Oct. 1995.

Y. Qin and S. Du, Line interactive UPS with the DSP based active power filter, in *Proceedings 17th International Telecommunications Energy Conference (IN-TELEC '95)*, The Hague, Netherlands, Oct.–Nov. 1995, pp. 421–425.

J. Rajagopalan and B. H. Cho, Space-vector modulated PWM converters for photo-voltaic interface applications: analysis, power management and control issues, in *Proceedings IEEE Applied Power Electronics Conference and Exposition (APEC '95)*, vol. *2*, Dallas, TX, Mar. 1995, pp. 814–820.

S. I. Moon, A. Keyhani, Estimation of induction machine parameters from standstill time-domain data, *IEEE Trans. Ind. Appli.*, vol. *30*, no. 6, Nov./Dec. 1994, pp. 1609–1705

J. Gutierrez-Vera, Use of renewable sources of energy in Mexico, *IEEE Trans. Energy Conversion*, vol. *9*, pp. 442–450, 1994.

J. Gutierrez-Vera, Modeling and analysis of a feedback control strategy for three-phase voltage-source utility interface systems, in *Proceedings 29th Annual Meeting of the IEEE Industry Applications (IAS '94)*, vol. *2*, Denver, CO, Oct. 1994, pp. 895–902.

J. Gutierrez-Vera, A single-phase voltage-source utility interface system for weak AC net- work applications, in *Proceedings IEEE Applied Power Electronics Conference and Exposition (APEC '94)*, vol. *1*, Orlando, FL, Feb. 1994, pp. 93–99.

M. C. Chandorkar, D. M. Divan, Y. Hu, and B. Banerjee, Novel architectures and control for distributed UPS systems, in *Proceedings IEEE Applied Power Electronics Conference and Exposition (APEC '94)*, vol. *2*, Orlando, FL, Feb.1994, pp. 683–689.

R. Uhrin and F. Profumo, Performance comparison of output power estimators used in AC/DC/AC converters, in *Proceedings 20th Annual Conference of the IEEE Industrial Electronics Society (IECON '94)*, vol. *1*, Bologna, Italy, Sept. 1994, pp. 344–348.

M. C. Chandorkar, D. M. Divan, and R. Adapa, Control of parallel connected inverters in standalone ac supply systems, *IEEE Transactions on Industrial Applications*, vol. *29*, pp. 136–143, 1993.

Jie Huang, Ching-Fang Lin, Internal model principle and robust control of nonlinear systems, *Proceedings of the 32nd IEEE Conference on Decision and Control*, vol. *2*, pp. 1501–1506, 1993.

A. Keyhani, H. Tsai, and T. Leksan, Maximum likelihood estimation of synchronous machine parameters from standstill time response data, *1993 IEEE Power Engineering Society Winter Meeting*, Columbus, OH. Jan. 31–Feb. 5, 1993.

N. Abdel-Rahim and J. E. Quaicoe, A single-phase delta-modulated inverter for UPS applications, *IEEE Trans. Ind. Electron.*, vol. *40*, no. 3, pp. 347–354, June 1993.

R. Cheung, L. Cheng, P. Yu, and R. Sotudeh, New line-interactive UPS systems with DSP-based active power-line conditioning, in *Proceedings 1996 IEEE Power Electronics Specialists Conference (PESC '96)*, vol. *2*, Baveno, Italy, June 1996, pp. 981–985.

J. S. Kim and S. K. Sul, New control scheme for AC–DC–AC converter without dc link electrolytic capacitor, in *Proceedings 1993 IEEE Power Electronics Specialists Conference (PESC '93)*, Seattle, WA, June 1993, pp. 300–306.

A. Keyhani and S. I., Moon, Maximum likelihood estimation of synchronous machine parameters and study of noise effect from DC flux decay data, *IEE Proceedings-C*, vol. *139*, no. 1, January 1992.

L. Morgan, P. D. Ziogas, and G. Joos, Design aspects of synchronous PWM rectifier–inverter systems under unbalanced input voltage conditions, *IEEE Trans. Ind. Appl.*, vol. *28*, no. 6, pp. 1286–1293, Nov./Dec. 1992.

G. J. Balas, J. C. Doyle, K. Glover, A. Packard, and R. Smith, *The μ Analysis and Synthesis Toolbox*, The Mathworks, Natick, MA, 1991.

A. Keyhani, Shangyou Hao, and Richard P. Schulz, Maximum likelihood estimation of generator stability constants using SSFR test data, *IEEE Trans. Energy Conversion*, vol. *6*, no. 1, March 1991.

A. Isidori and C. I. Byrnes, Output regulation of nonlinear systems, *IEEE Trans. Automat. Control*, vol. *35*, no. 2, pp. 131–140, 1990.

J. T. Boys and P. G. Handley, Harmonic analysis of space vector modulated PWM waveforms, *IEE Proc.*, vol. *137*, pt. B, no. 4, pp. 197–204, July 1990.

S. Hara, Y. Yamammoto, T. Omata, and M. Nakano, Repetitive control system: A new type servo system for periodic exogenous signals, *IEEE Trans. Autom. Control*, vol. *33*, no. 7, pp. 659–666, 1988.

M. Mansour, F. Kraus, and B.D.O. Anderson, Strong Kharitonov theorem for discrete systems, in *Proceedings of the 27th IEEE Conference on Decision and Control*, vol. *1*, 1988, pp. 106–111.

S. Rahman, R. Bhatnagar, An expert system based algorithm for short term load forecast, *IEEE Trans. Power Syst.*, vol. *3*, issue 2, pp. 392–399, May 1988.

I. Dobson, H.-D. Chiang, J. S. Thorp, and L. Fekih-Ahmed, A model of voltage collapse in electric power systems, in *Proceedings of IEEE Conference on Decision and Control*, 1988, pp. 2104–2109.

T. Haneyoshi, A. Kawamura, and R. G. Hoft, Waveform compensation of PWM inverter with cyclic fluctuating loads, *IEEE Trans. Ind. Appl.*, vol. *24*, no. 4, pp. 659–666, July/Aug. 1988.

R. J. Thomas, A. G. Phadke, and C. Pottle, Operational characteristcs of a large wind-farm utility system with a controllable AC/DC/AC interface, *IEEE Trans. Power Syst.*, vol. *3*, no. 1, pp. 220–225, Feb. 1988.

H. W. Van Der Broeck, H. Skudelny, and G. V. Stanke, Analysis and realization of a pulsewidth modulator based on voltage space vectors, *IEEE Trans. Ind. Appl.*, vol. *24*, no. 1, pp. 142–150, Jan./Feb. 1988.

E. J. Davison and B. Scherzinger, Perfect control of the robust servomechanism problem, *IEEE Trans. Autom. Control*, vol. *32*, no. 8, pp. 689–702, 1987.

K. C. Kalaitzakis and G. J. Vachtsevanos, On the control and stability of grid connected photovoltaic sources, *IEEE Trans. Energy Conversion*, vol. *EC-2*, no. 4, pp. 556–562, Dec. 1987.

A. Kawamura, T. Haneyoshi, and R. G. Hoft, Deadbeat controlled PWM inverter with parameter estimation using only voltage sensor, *Conference Record of IEEE Power Elec. Spec. Conf.*, 345–350, 1986.

A. Kawamura and R. G. Hoft, Instantaneous feed back controlled PWM inverter with adaptive hysteresis, *IEEE Trans. Ind. Appl.*, vol. *IA-20*, no. 4, pp. 769–75, 1984.

T. S. Key, Evaluation of grid-connected inverter power systems: the utility interface, *IEEE Trans. Ind. Appl.*, vol. *IA-20*, no. 4, pp. 735–741, July/Aug. 1984.

B. K. Bose and H. A. Sutherland, A high performance pulse-width modulator for an inverter-fed drive system using a microcomputer, in *Conference Record of IEEE IAS Annual Meeting*, 1982, pp. 847–853.

P. Zoigas, Delta modulation technique in static PWM inverters, *IEEE Trans. Ind. Appl.*, vol. *IA-17*, pp. 289–295, 1981.

T. Inoue, M. Nakano, and S. Iwai, High accuracy control of servomechanism for repeated contouring, in *Proceedings of 10th Annual Symposium on Incremental Motion Control Systems Devices*, 1981, pp. 258–92.

T. Inoue and M. Nakano, High accuracy control of a proton synchroton magnet power supply, *IFAC*, vol. *20*, pp. 216–221, 1981.

E. Davison and I. Ferguson, The design of controllers for the multivariable robust servomechanism problem using parameter optimization methods, *IEEE Trans. Autom. Control*, vol. *26*, issue 1, 1981

P. Wood, *Switching Power Converters*, Van Nostrand Reinhold, New York, 1981.

K. Zhou and J. C. Doyle, *Essentials of Robust Control*. Prentice Hall, Upper Saddle River, NJ, 1981.

K. J. Astrom, Maximum likelihood and prediction error methods, *Automatica*, vol. *16*, pp. 551–574, 1980.

B. A. Francis and W. M. Wonham, The internal model principle for linear multivariable regulators, *Appl. Math. Optim.*, vol. *2*, no. 2, pp. 170–194, 1975.

INDEX

Integration of Green and Renewable Energy in Electric Power Systems. By A. Keyhani, M. N. Marwali, and M. Dai
Copyright © 2010 John Wiley & Sons, Inc.

Printed and bound by CPI Group (UK) Ltd, Croydon, CR0 4YY

16/04/2025

14658594-0004